# *Wild Brothers:* MAINE ANIMAL TALES

# *Wild Brothers:* MAINE ANIMAL TALES

JACK ALEY

ILLUSTRATED BY
MARGARET B. CAMPBELL

**LANCE TAPLEY, PUBLISHER**
Augusta, Maine

Text copyright ©1987 by Jack Aley
Illustration copyright ©1987 by Margaret Campbell

All rights reserved

The author wishes to thank many people for their help, particularly Bud Leavitt of the *Bangor Daily News,* Gene Letourneau of the Guy Gannett Publishing Co., Dick Anderson, John Hunt, and Linwood Rideout.

Some material from "The Magnificent Ghost" originally appeared in an article by Herbert Adams in *Habitat: Journal of the Maine Audubon Society,* August 1985. Appreciation is extended for its use.

"Damn Mussels" originally appeared in slightly different form in the *Maine Times,* and is reprinted here with permission.

Information for "Bruno And Ursula" came from *Wild Brother,* by William L. Underwood, published by Atlantic Monthly Press, Boston, 1921.

No part of this book may be reproduced or transmitted in any form or by any means, electronic or mechanical, including photocopying, recording, or by any information storage and retrieval system, without the written permission of the publisher, except by a reviewer quoting brief passages in a magazine, newspaper or broadcast. Address inquiries to Lance Tapley, Publisher, P.O. Box 2439, Augusta, Maine 04330.

Printed in the United States of America

## CONTENTS

| | |
|---|---|
| *Bully And Mike* | 2 |
| *The Harvard Varsity Club Bears* | 10 |
| *Bruno And Ursula* | 20 |
| *Road Kill* | 28 |
| *Love For A Skunk* | 36 |
| *No Lamb* | 42 |
| *Trout Fever* | 50 |
| *Maine Man Meets Mermaid* | 60 |
| *Thanksgiving Rats* | 68 |
| *Most Efficient Killer* | 76 |
| *Sexing The Ducks* | 86 |
| *Does A Bear . . . ?* | 94 |
| *Damn Mussels* | 104 |
| *The Dangerous Sport* | 112 |
| *Twenty-One Days In The Woods* | 122 |
| *Pinned* | 132 |
| *The Magnificent Ghost* | 140 |
| *Caribou: The Evolution* | 148 |

# Bully
# and
# Mike

MIKE ROBINSON awoke and smelled bacon and eggs, which on a cold November morning signaled one thing: His father, normally a toast-and-coffee man, always ate a full breakfast when they were going hunting.

He pulled himself out of bed and started to dress warmly. He looked out his bedroom window to the town—and to a vast man-made cloud above it. Isolated in Maine's commercial forest, Millinocket had been built to transform trees into paper.

Mike, eighteen, had an entry-level job in the paper mill. His father, Wilmot—or "Wiggie," as everybody called him—was a foreman. They looked a lot alike, both short and dark, and shared a passion for the outdoors. Wiggie had been a registered guide for twenty years. Mike had just become one at the minimum age.

They hunted or trapped or fished at every opportunity and had arranged their schedules at the mill so they had their days off together. On an average of one day in five,

Wiggie and Mike Robinson struck off into the ocean of woods surrounding their town.

Wiggie was proud of his son's savvy in the woods. Mike had been a quick study with a shotgun and a proficient bird hunter as a young boy. He had become an expert fly fisherman in one season. Wiggie figured he had taught Mike everything he could teach him. Now, the teaching over, the father just enjoyed his son's companionship.

After their big breakfast, father and son loaded their fourteen-foot canoe on top of the old Volkswagen bus parked in front of their Medway Street home. They put in it their cleaned and oiled rifles, Wiggie the 30/30 he'd owned since 1939 and Mike his .32 special. For lunch there was a roast beef sandwich for the father and a big grinder for the son.

At the border of the town they got on the unpaved Golden Road, a Great Northern Paper Company logging highway (so named for its expense). The first several miles of the Golden Road heading north out of Millinocket was a wondrous sight, a two- to three-story canyon of wood, a huge reserve of stacked trees. Wiggie had always viewed the enormous stockpile with satisfaction. It represented job security for him, and now for his son.

Wiggie Robinson nursed the old bus along the rough wide road for about an hour and a half, the time it took to travel the sixty miles to their destination. He watched carefully for the familiar landmarks, and then pulled over.

They were near the shore of Salmon Pond, a small lake at the foot of Salmon Mountain. They carried their canoe to the shore of the pond, loaded it with gear, and paddled across. As far as they were from civilization, Wiggie still wanted the buffer of the pond between him and the road.

Wiggie had hunted in this area before. It had been cut

heavily—brutally, in some estimations—but the heavy cutting released new vegetation that favored wildlife. Pioneer growth such as raspberries, pin cherry, poplar, and soft maple provided tender browse. This was good deer country.

Wiggie and his son conferred briefly at the far shore of Salmon Pond and agreed to meet at the canoe at noon. They used to hunt together but realized that at day's end they had only one story to tell.

Wiggie walked quickly up the wide tote road that headed up a flank of Salmon Mountain. It was a brisk, windy day marked by high, fast-flying clouds. A succession of hard November frosts followed by strong winds had stripped most of the leaves from the trees. There was a soft cushion underfoot.

Beech leaves hang onto their silvery branches longest, and he was climbing into a beech ridge when he heard the leaves rustling. His eyes were drawn to the sound. The moose blended in almost perfectly beneath a burnished brown canopy of young beech.

Seeing a moose in the Maine wilds was a rare enough thing, even for a man like Wiggie who was often in the woods. It was possible, in the early seventies, to spend a week on the Saint John River, Maine's wildest, and not see one moose.

The animal was making a comeback, but in November of 1970 few people imagined it would become so populous as to be hunted again. It wasn't until 1980 that a moose-hunting season was instituted.

To Wiggie, the huge animal had always symbolized the wild, and any encounter with one was a special treat.

He stood transfixed by the large cow. She moved slightly and his attention was diverted. Less than twenty yards away from her were three more moose, two big bulls and a spring calf. The little fellow, as Wiggie remembered, was "standing sound asleep in a spot of sunshine."

Moving lightly and almost soundlessly, he put his gun against a tree and worked his way slowly up the slope toward the calf. He had never been so close to such an animal before. As the man started to move, the other three moose ambled away.

The calf was still dozing when he came to within an arm's length. He poked the little moose gently in the ribs and jumped back. The calf opened its eyes slightly, raised its head, and fell back to sleep.

"I was thrilled to be so close to a wild animal like a moose," he remembered. "I had to touch it."

Feeling braver, he started stroking the animal as he would one of his dogs or one of his daughter's ponies. The moose opened its eyes but didn't move. He worked his hand up the calf's back and scratched behind its ears. Then he started rubbing its belly. The moose, which he instantly named Bully, loved it. Wiggie loved it, too, and he continued patting, scratching, and soothing Bully for about ten minutes.

He thought to himself, "Wouldn't I like Mike to see this moose." He sensed that this encounter was a once-in-a-lifetime experience and he had to share it with his son.

He had a piece of rope with him in case he shot a deer and needed to haul it out. He formed a halter with a bowline knot and gently placed it over Bully's neck. He tugged. Bully responded. It was firmly set in Wiggie Robinson's mind to lead Bully the three-quarters of a mile back to the canoe.

There were blowdowns on the slope of Salmon Mountain and Bully balked like a mule at the first one. Try as he might, he couldn't get the animal to step over the fallen trees. He remembered something his daughter had once told him. If a horse won't move, she said, pull as hard as you can and then let go.

He did so and Bully landed on his rump, startled. The next time he pulled, Bully jumped right over the blowdown—and every subsequent one on the quarter-mile trip back to the tote road.

Once on the woods road, Wiggie tied the loose end of the rope to his waist and walked along in front of Bully, who paused from time to time to browse. The man figured that if a warden ran across Bully and him, he would be arrested for molesting a wild animal, but abuse of Bully was the farthest thing from his mind.

Back at the pond, he tied Bully to a stump and made his way to the canoe where Mike was eating his ham and cheese grinder.

"Mike, I've got something to show you," he said.

"You got a deer, Dad! You got a deer!" Mike exclaimed.

"No, no, I've just got something I want to show you. Leave your gun here and come along with me."

Wiggie Robinson wanted to surprise his son, so on the short walk he kept Mike's attention away from the pond. They were within fifty yards of the tethered animal when Mike spied Bully. He ran ahead a few steps and cried, "Hey, Dad, look, there's a moose!"

Mike was wearing a blaze orange vest and when Bully got an eyeful of that the moose pulled back and broke the rope.

"Mike, take that vest off," Wiggie said. "It might be alarming him. Now, come on over here. I want you to meet my friend Bully."

Mike shrugged off his vest. He watched transfixed as

his father slowly approached the young moose—who settled down quickly— and took the rope.

The next twenty minutes were like a dream. Father and son scratched Bully and patted the calf and poked its ribs to see how much meat was on them. Mike joked that maybe they should take Bully home so it could play with his sister's horses. They almost had to pinch themselves to realize they were gamboling with a moose in the Maine wilderness.

It was an experience that had to be fleeting. Soon, Wiggie smiled and said it was time. He took off the halter and slapped Bully on the rump. The young moose slowly walked away, looking back frequently as if wondering when its new friends would follow.

There was no question of hunting anymore that day. They packed their gear quickly, paddled back across Salmon Pond, and drove home, talking about Bully all the way.

"They came in bubbling all over that night," Joyce Robinson recalled. "They told me the story and I couldn't believe it. But it happened."

EARLY ONE MORNING four years later, Wiggie and Mike Robinson were close to finishing their shift in the paper room. Wiggie was still a foreman. Mike, who was now a husband and a father, had moved up two notches in the paper room to a third hand, a winder operator. Father and son still took their time off together. They were going hunting in two days' time.

It was about six-thirty a.m. and Mike was attending to the winder, which cuts large rolls of paper to customers' specifications. He was making sure that it was operating properly, that the slitters were cutting, and that there were no corrugations. He was protected from the winder by a gate.

Behind him was a levelator, a huge set of steel arms that when manually activated rises out of the floor, moves to the bed of the winder, and takes off the giant rolls of paper. There was no protecting gate for the levelator and Mike had his back to it.

"I was foreman that night," Wiggie recalled. "I was in the back end of the paper room and I saw people running with a stretcher. Being in charge, I took off. I got closer and I saw the looks in the other men's faces. They tried to stop me. They didn't."

Wiggie Robinson says he has always had a problem with the death certificate. It said human error caused his son to be crushed to death by the steel arms of the levelator. But he says there was nobody near the switches that activated the machine.


IT WAS MANY MONTHS before Wiggie could go back to the woods. But one day he threw the canoe on the VW bus and drove up alone. He saw the canyon of trees differently now. He was determined to take early retirement from the paper company.

When he got to Salmon Pond, he unloaded the canoe and paddled across. He had no gun and no food. He found the stump within minutes. The rope was still attached. He placed Mike's orange vest on the stump and sat down. He closed his eyes and saw Bully and his son playing.

He watched them for a long time. He has not been back to Salmon Mountain since.

# THE HARVARD VARSITY CLUB BEARS

MONTGOMERY MACMILLAN moaned as he gazed out the train window at his beloved Boston. He hated what he was about to do. He detested where he was about to go—which was, in his mind, nowhere.

He had resisted making a hunting trip into the Maine woods for many years. His half-brother Tom had always been insistent on Monty making a trip with him, but had never prevailed until now. Tom, an outdoors enthusiast, had hunted and fished all over Maine all his life and owned land on Moosehead Lake.

It took the occasion of Tom MacMillan's fortieth birthday to persuade Monty MacMillan to make his maiden voyage into the wild.

"Monty, the only gift I want from you on my fortieth is to come with me to Moosehead on a deer-hunting trip," Tom said. "I want you to experience this just once. I'll never ask you to go again if you'll come this one time."

Monty was trapped. It was, after all, a big event in Tom's life, and he loved his younger brother deeply, despite their enormous differences. "Okay, okay, I'll go," Monty sighed. "But I know you, Tom, and your damned Dartmouth outdoors heritage. Don't try to impress me. Please make this a trip for a beginner or I'll never forgive you."

Monty MacMillan was an unimposing figure in his first-class seat on the *State of Maine* as it pulled out of Boston. He was a dumpling of a man in his early fifties. His head was nearly bald and his skin chalky white, the result of a life cheerfully led indoors.

His expensive silver-rimmed glasses were perched on a large, ruddy nose that bespoke its owner's affinity for twelve-year-old unblended scotch. He was reading a novel by Henry James. It was Monty's third time through his favorite author's body of work. He'd majored in English at Harvard and had toyed with the idea of a career in teaching until he fully understood the financial consequences of pursuing one. He'd gone into the insurance business instead and become a reasonably rich man.

In the late 1930s, Boston quickly gave out to relatively unspoiled landscapes, and the ratio of trees to settlements soon made Monty uneasy. He seriously considered changing course in Portland, a frontier town in his mind, and returning to Boston on the next train. But he knew that he could not disappoint his brother. He'd just have to get this ridiculous business out of the way.

He changed trains in Portland and the slower, older train into the north country symbolized for him the raw, backward circumstances into which he was heading. He was dumbfounded by the scope of the wilderness that surrounded him on both sides and might, he feared, swallow him up whole. Normally, Monty started drinking at exactly 5:30 p.m. Today, he unscrewed his well-worn silver hip flask at noon.

It was midafternoon on a Friday in late November when his train pulled into Greenville. The weather, twenty degrees colder than in Boston, suggested snow. He huddled inside his new canvas hunting coat, part of the outfit his brother had bought for him at L.L. Bean. He felt, and sensed, that he looked absolutely ridiculous.

"You Mr. MacMillan?" the tall, thin man asked in a heavy Maine accent. "Your brother described you just right. I'm Joe Greene, one of the guides. Come on. The truck is over there."

Greene guided Monty to one of the most dilapidated vehicles the Boston insurance man had ever seen. The ride to the camp took an hour over rough roads. The truck's heater didn't work and the laconic guide grunted one-syllable responses to a couple of questions politely put forward by Monty. The rest of the ride was made in silence.

The road didn't go all the way to the lakeside camp and Monty had to tote his gear, neatly packed into his new Bean duffle, about a half mile. This walk represented the most exercise he had had in twenty years. He arrived at the cabin exhausted. It was three-thirty in the afternoon, too late to start hunting, which was fine with him, and too early to eat camp food, another wilderness prospect he had been dreading.

So Monty crawled into his bunk bed with his twelve-year-old unblended scotch and his Henry James. He drank and read until sleep overwhelmed him. The only comforting thought in this godawful place was the prospect of oblivion. Thank God he'd brought plenty of whiskey.

The next thing he was conscious of was the sound of his brother's voice. Tom had tried to wake Monty for dinner the night before but couldn't penetrate his sodden slumber.

"For God's sake, what time is it?" Monty groaned. His

head was in a vise and his eyeballs felt congealed. There was an empty scotch bottle on the rough pine table next to his bunk.

"It's five in the morning. Breakfast is ready. You've got to get moving. The guides will be ready for us in an hour," Tom said.

"Not for me, they won't," Monty replied. "There's no way I'm going out there today. I feel horrible. I'm too sick to hunt. I'm too sick to do anything but stay right here—where I might add, I don't want to be."

"Damn it, Monty, you promised me. This is my birthday present from you, remember? I can't go back in there and face the boys and tell them my brother won't get out of bed. Just make an appearance. It's just this once. I know you won't be back."

Monty struggled into hunting clothes so new they squeaked. He looked out the door. "God, it's still night out there!" He hadn't been up so early since childhood Christmas mornings. There was a porch light at the mess hall and he stumbled in that direction.

The smell of bacon that greeted him when he opened the door, given his delicate condition, almost made him throw up. He took an empty seat next to his brother and tried to make small talk while forcing down some boiled coffee. He ignored his food. He thought he might be dying.

The guide, Joe Greene, was waiting for him when he came out of the mess hall. Greene had been prepared by Tom MacMillan for the level of his brother's prowess as a hunter.

Under the porch light, Greene patiently showed Monty how to load his five-round carbine and then instructed him how to fire it. Monty had never shot nor loaded a real hunting rifle. He nodded grimly as Greene went through the ritual.

"There's enough light now, Mr. MacMillan. Let's get going," Greene said.

Monty followed his guide across the clearing in front of the camp for about a hundred yards, and then he stopped. "Mr. Greene, I appreciate all your trouble but I'm really quite sick and I'm going to head back to bed. Here's a ten-dollar tip. Please don't tell my brother."

The guide refused the proffered note with a brief wave of his hand.

"Can't take that, Mr. MacMillan. Been hunting with your brother for years. Couldn't lie to him. Just head up that tote road a ways and give it a little time. The air will make you feel better if you've got what I think you've got. I'll take the fork here to the main road."

Monty did what he was told. He walked to the first bend in the woods road and soon was surprised to feel slightly better. He waited around awhile until he was sure the guide had covered some distance. He started to head back to camp and his bunk.

When he turned around, the bear was in the middle of the road. It wasn't more than twenty yards away from him. Monty had never seen a live bear before, not even in a zoo because he had never, in his memory, been to a zoo.

His brother had told him that if you see a bear in the woods, shoot it. And this bear was blocking Monty MacMillan's path back to his warm bunk. So he quickly reviewed what Greene had told him and shot the bear dead in the side of the head.

Warily, he walked up to the bear and as he did he heard a rustle in some beech trees next to the road. He looked up and saw two cubs. They weren't any distance away at all. He shot and killed them both.

Suddenly, Monty realized that he felt acutely alive. He'd heard the jocks at the Harvard Varsity Club talk about adrenalin surges when recounting their athletic exploits. This must be something like what they were talking about.

At this point an experienced hunter would have dealt with the situation at hand and stayed put. Instead, Monty excitedly headed through the woods in an attempt to reach his guide walking on the other road. Joe Greene would know what to do with three dead bear.

He was less than a hundred yards off the road when he moved across a clearing in the woods and came upon another bear. He shot it in the head just as he had the one on the road. "Almost by habit," as he later recounted the incident, he looked up and saw a cub in a nearby tree.

Bang. A fifth bear in five shots was dead at the hands of Montgomery MacMillan, Boston insurance executive.

Nobody back at the camp was exactly clear what the record was before. Nobody doubted that Monty had set a new one—not only for the most bears killed in one day but for the period of time it took him to do it—one hour and twenty minutes from the time he left camp.

MONTY HAD ALWAYS BEEN a terrible misfit at the Harvard Varsity Club. He dearly wanted to belong, as did most Harvard men, but that required earning an athletic letter. During his senior year he had schemed to become manager of the varsity baseball team—and so earned his "athletic" letter.

Over the years, he went to the club often and was intrigued by the blend of elegance and machismo. But whenever talk turned to sports, as it often did, Monty felt terribly inadequate. He had been, after all, a Milquetoasty undergraduate.

So he knew almost immediately what to do with his five bear. They were going home with him. The guides gutted the animals. That was all. He wanted the skins and he wanted the meat. His brother Tom was mystified. "Monty, what the hell are you going to do with bear meat in Boston?"

Monty survived the rest of the hunting weekend in better shape than he'd ever imagined. He saw no more bear and never spied a deer and never fired another shot. That made no difference. He couldn't wait to get back home.

He entrained at Greenville early Monday morning and was home that evening, delighted to be back in civilization. He had a hot bath, a good meal, and gave his wife a brief account of his trip.

At his office the next morning, he made two calls before he started his day's work. The first was to the best butcher in Boston. The second was to the Harvard Varsity Club, which he asked to reserve the Friday night before Christmas. He invited everyone he knew who was a member of the club. And he invited his brother Tom.

Everybody agreed there had never been an affair quite like it at the club: a five-course meal with one hundred and thirty-five pounds of bear steaks as the main course. Monty had chosen a heavy claret to go with the bear meat, which had marinated for a day in a sauce the ingredients for which Monty had researched meticulously. He'd insisted that the meat be grilled over cherry-wood coals.

Nobody knew he had eaten bear meat until after the dinner was over and Monty informed his guests in a brief and enigmatic talk. "I just wanted to invite you to a wild dinner in the most civilized club in the world," he told them, and then retired to stand before the fire with his cigar and brandy. He guessed what was coming and he had prepared for it.

"Now, Monty," asked a tall, muscular fellow who'd been an all-Ivy-League end, "just where in hell did you get that bear meat?" Monty said in a flat voice that he had killed five bear with five shots in Maine and brought the meat home with him and wasn't it delicious?

"Yes, it was great," said the incredulous former athlete. "But you are full of it. We all know you never leave this town. You're no hunter. Did your brother set this up for you?"

By this time, a little crowd had gathered around Monty. He had never, ever drawn a crowd there before. Smiling ever so slightly, Monty took a small leather folder with gilded edges out of his coat pocket.

Under his hunting license was a formal documentation of his record-setting bear slaughter in Maine. It was signed by the director of the Maine Fish and Game Department and by the governor of Maine. Monty had received it by registered mail only two days before the dinner.

He handed it to the former football player who read it slowly and shook his head and passed it on. The others read it also, and, when they were done, they raised their glasses in a spontaneous toast to Montgomery MacMillan, former misfit at the Harvard Varsity Club.

# Bruno
# and
# Ursula

CARENCE WELDON AND HER FAMILY had come in with the snows, and they would leave when the snows melted. They were twenty-five miles from Schoodic Lake and only a rough one-way logging road connected the camp with civilization.

Carence, dark and slim, looked out the small window of their cabin. The men had finished their noon meal of pork and beans and were grinding their axes in preparation for the afternoon's cutting. They were talking quietly among themselves and chewing tobacco. Shortly before sunset they would return with icicles on their beards and one thought in their minds: food—especially, red meat.

Carence Weldon's husband, William, was the camp cook. Mr. Gordon, the owner of the logging camp, had come to their small Maine village in the fall and offered William the job; her husband had a good reputation and backwoods cooks were always in demand. Weldon had declined. He didn't want to leave his pregnant wife and three children for the entire winter. Mr. Gordon then offered them the log cabin next to the men's quarters.

Weldon had consulted with his wife. They would be a long way from a doctor when it came time to give birth in the winter. Women almost never came to the camps. Except for the company of her children, it would be a lonely several months. William would put in long days. She knew, however, that her husband loved the life in the logging camp, and she had a strong sense of adventure. So they had travelled down the long logging road in late November. They didn't plan to emerge from the wilderness camp until March at the earliest.

The biggest challenge to William Weldon in his job of feeding the camp was offering a variety of meat. Pork was the foundation of each meal. Add to that the ingredients for baked beans, saleratus biscuits, molasses ginger cookies, and applesauce. That's what he was provided to work with in the camp kitchen. Anything else, he had to get himself.

Well, there was deer, and Weldon knew what to do with venison. The men joked about the occasional "deer attacks" in the woods, and they carried their guns to "defend" themselves against the deer when they "charged." It was very illegal to shoot them out of season in 1903, but that was routinely overlooked.

Weather permitting, Weldon snowshoed down to a nearby lake and ice-fished. He hacked holes through two feet of ice and fixed his lines to alder poles. On good days, he stacked up salmon and trout like firewood. It took quite a stack of fish to feed the camp, but he found his reward in the intense, appreciative hush at dinnertime.

One night in mid-January, the swampers came back to camp with a story. They had been clearing away brush and timber for a new tote road. Every time they got near a spot on a small ridge the horses started acting up. They reared and snorted and plunged. The men couldn't figure it out.

William Weldon listened to them talk. He was older than most of the others and had spent all of his life in the woods. There was a bear nearby, he said. The men laughed at him. It was twenty below outside. There was four feet of snow on the ground. There's no bear around in weather like that, they said.

The next morning after breakfast, Weldon snowshoed to the ridge. He quickly found what he was looking for at the base of a huge pine: a small hole rimmed with ice and frost.

He started digging with his snowshoes and, when he'd dug to the ground, he found a large burrow into the roots of the tree. He cut a sapling and poked it into the hole. He hit something soft and yielding and jumped back at the deep warning growl.

He snowshoed back to camp for his gun and some help. He returned and poked the barrel of his rifle into the hole and fired twice. It took another man and Weldon a half-hour to wrestle the two-hundred-pound dead sow out of her den. They skinned the bear on the spot and began preparing the meat. It had been a long time since he'd served up bear meat to a camp of hungry loggers. They were in for a treat. Besides, the bear skin would fetch twenty-five dollars and the bounty ten.

The cry was eerily human and stopped Weldon cold as he was cutting into the sow's flank. He lay down flat on the snow and felt around in the den with his bare hand. The cub that he drew out of the burrow was small and furry and warm. It was no bigger than a small squirrel, and its eyes were still closed. Weldon slipped the tiny creature into the pocket of his overcoat.

His children marvelled. Christmas had been lean and the bear was like a gift from heaven. The lumbermen, returning from work, stopped to gaze at the cub. They touched it tenderly with their rough hands. Many of them had not seen their own children for months. The cub was the talk of the camp.

The Weldons' baby girl had arrived in late December. Fortunately, the birth had been routine. Carence Weldon always had trouble naming her children; she still had not given her baby a name.

Carence was breast-feeding when William brought the cub into their cabin. She felt the atmosphere of the place change instantly with the presence of the tiny and appealing animal. She felt drawn to it and put down her baby to prepare a deerskin-lined box for the cub.

Soon, the cub started crying and whimpering. The message was universal. Bruno, as one of the children had named it, was hungry. There was no cow in the camp, so there was no cow's milk. The men drank tea and coffee. Canned and evaporated milk had not reached so far into the wilderness. Death by starvation seemed inevitable for the orphaned cub.

Carence Weldon did not have to hear the cub crying for long before her maternal instincts found a solution. She told William what she planned to do and he said no wife of his was ever going to do a thing like that.

"You killed the mother," Carence told her husband. "And I alone can save the cub. Poor, lonely creature. I'll do my best to bring it up."

William Weldon never argued for long when his wife had iron in her voice. She was indomitable when she got like that. So he accepted the inevitable. His wife was going to nurse a wild animal.

Carence started nursing Bruno daily on January 23, 1903. Bruno gravitated to the left breast, which was fine because the little Weldon girl had always preferred her mother's right breast. In many ways, Bruno and the baby were twins. They even seemed to get hungry at the same time. Carence would rock away winter days in front of the fire with her bear sucking on one breast and her baby daughter on the other.

Word of this highly unusual situation spread rapidly to nearby camps, and then to the nearby towns in Maine, and then beyond—even to Boston. A Bostonian who had a summer place at Schoodic Lake made a trip to the logging camp to witness it for himself.

This Mr. Underwood, who fancied himself a writer, spent several days in the camp documenting what had happened. He brought along a camera and persuaded Carence to pose while feeding Bruno and the baby. She felt indebted to the educated man, for he had come up with a name for her little girl: Ursula, Latin for a she-bear.

Mr. Underwood realized the commercial potential for taking a wild animal back to civilization who had been saved and suckled by a human female. He offered Carence Weldon twenty-five dollars for Bruno.

"Sell my cub?" she huffed. "I guess not! Why, Mr. Underwood, you haven't enough money to buy him. I wouldn't any more sell Bruno than I would sell my baby." So Mr. Underwood went back to Boston without the bear but with enough material for a little book, which was published some years later.

Meanwhile, Mr. Gordon, the camp owner, also understood the bear's commercial potential and made a special trip out to the camp to get Bruno. He was the baronial type who controlled everything that happened at his camp, and had agreed to sell the cub without consulting with the Weldons.

Carence was devastated when Mr. Gordon demanded the bear. She refused at first and pleaded with her husband to intervene. William tried and failed. Gordon would take the bear and do so that evening, leaving it in the men's quarters overnight.

Sound travels clearly through intensely cold, still air. That night it was thirty below zero and the cries of Bruno reached Carence easily and touched her deeply. She sobbed through the night. The men who'd been forced by Gordon to take in the cub fared no better. Bruno's cries not only kept them awake, but stirred something deep within them.

Many of these rough lumbermen regarded Bruno as they would one of their own children. They loved to play with the cub after a long day in the woods. His softness and playfulness took a harsh edge off the life in the camp.

The next morning the men called what they termed an "indignation meeting." Their vote, after very little discussion, was unanimous. Mr. Gordon was called into their quarters. The foreman was the spokesman.

"That woman saved the cub's life," the foreman said. "You must give it back to her or you can stay here and cut your spruce alone."

That day, Carence Weldon was back to feeding her cub. That night, the men had salmon and trout for dinner.

# ROAD KILL

"PEOPLE ARE ALWAYS TALKING about tainted meat. There's no such thing as tainted meat! As long as you can cook it, you can eat it. I've eaten deer all blown up. I've stuck a knife into them and the air hisses out and then I've eaten them. That doesn't bother me. Tainted meat . . . that's an old wives' tale."

Richard Anderson, formerly Maine's conservation commissioner, is sipping a draft beer in a Freeport bar just a few yards from L.L. Bean, "the store that knows the outdoors." Anderson is talking of an outdoors beyond the ken of L.L. Bean.

"One spring morning about twenty years ago, I was driving my pickup along Route 302 in Windham and, lo and behold, I saw a dead deer on the side of the road. It was the only deer road kill I've ever discovered. I've never even hit a deer.

"I jammed on the brakes and threw the deer, about a hundred-and-thirty-pound doe, in the back of my truck. I said to myself, 'Gosh, I've got to call a warden.' So I stopped at the first phone booth I could find, which happened to be at an electric company. I parked the truck and walked to the phone booth and called Raymond Curtis, the nearest warden.

"I was telling him the story when I happened to look out the window of the phone booth toward the truck. You know how when you take a leak on blacktop and it pools and then runs? Well, the deer had started bleeding and there was a small fall of blood from the truck. It had pooled and was coursing over the blacktop to the front door of the shop.

"I couldn't get Curtis to stop talking and I had to be accommodating because I really wanted that deer meat. He kept talking and I watched the pool of blood deepen and expand in front of the door of the electric company. There were customers in the shop, and all I could think about was the reaction of the first customer who opened the door to leave. I finally got Curtis off the phone and got out of there fast."

Dick Anderson has had a lifelong obsession with carrion. It is as natural for the wiry, fiftyish conservationist to jam on his brakes for a road kill as it is for other people to jam on theirs at a scenic overlook.

"I have had a reputation for thirty years for picking up road kills. I guess it started when I worked for the Maine Fish and Game Department as an assistant regional biologist. I couldn't stand to see all the road kills go to waste."

Over the years, Anderson has consumed most of the species of free protein Maine's roadside delicatessen serves up. His life list includes porcupine, raccoon, squirrel, muskrat, deer, beaver, and partridge. And, among domesticated animals, one black chicken.

It was a hot day in the summer of 1983 and Commissioner Anderson was heading down the superhighway with, in the back seat, Jeff Pidot, then chief of the Maine Land Use Regulation Commission.

"There, right in the middle of I-95 around Burnham, was a black chicken. I hit the brakes, ran across the interstate, got the chicken, ran back, and threw it in the back seat right on Jeff Pidot's lap. Well, Jeff is a Philadelphia lawyer and he just about had a heart attack."

Pidot, now a Maine assistant attorney general, will never forget the incident. "Oh, my God. I'm a very conservative, quiet guy. So what do you do, what is your reaction when you are sitting in the back of a car heading down I-95 with your boss driving and he screeches to a halt and tosses a dead animal into your lap? I was blown away, I guess."

Anderson took the bird home and, when cleaning it, found an egg inside, which he washed and saved for breakfast. He and his wife, Patricia, dined on road kill that evening. There were leftovers.

"The next morning, Patricia made up a chicken sandwich and put a label on it that read 'Black Chicken Sandwich.' I brought it to Jeff at work. I didn't think he would, but he ate it," said Anderson.

Dick Anderson grew up in Brockton, a reasonably civilized town in Massachusetts where people generally don't go around scooping up dead meat off the pavement for their dinner tables. He was drawn to wilder country early on. He got his degree in wildlife management at the University of Maine at Orono and then stayed in the state. He headed the Maine Audubon Society from 1970 to 1977 and then went on to run a waste recycling company. He was appointed conservation commissioner in 1981 and served until a Republican became governor in 1987.

The conservation establishment has been good to Anderson. He has never needed to eat road kill as a nutritional supplement.

Nor has he had to trap for money. But he has had a trapping (and hunting and fishing) license all his adult life and he will take a pelt anywhere he can find one.

"Last November, I was headed up the interstate to a meeting in Caribou. It was about six-thirty a.m. and I was just north of Bangor. I saw an animal. It was really squashed. 'Holy mackerel,' I said to myself, 'that was a fisher.' I had never seen a fisher road kill before.

"I'd gone half a mile when I said to myself, 'Dick, you gotta go back and get that.' I slammed on the brakes and backed up.

"It was squashed, all right. I mean really squashed. There was no dinner in it. Imagine an emptied toothpaste container. But lots of times when animals get squashed their skins are okay. I figured the skin was worth about seventy-five dollars, so I took it home and worked on it. It only had a little rip. I stretched it and really did a nice job. It was beautiful. I was absolutely convinced I was going to get seventy-five bucks for it.

"I knew I had to get the skin tagged by the fish and wildlife people, but I didn't bother to bring it in right away. I waited until I had trapped some beaver to bring in with it. Finally, I went into the fish and wildlife office in Gray.

"The warden said, 'Dick, I'm afraid we're going to have to take this fisher skin away from you. You have to have something like this tagged within ten days of the end of the season.'"

Anderson says, "I guess I felt all right about it after I got over the initial disappointment. The pelt didn't go to waste. Fish and wildlife has auctions. I figured it was like making a seventy-five-dollar contribution to the department."

PATRICIA ANDERSON comes into the bar to pick up her husband. She fits easily into the conversation. The Andersons obviously have teamed up on people before. On the subject of road kills, they are puckish. He chortles a lot. Patricia wears a smile of resignation.

He says he has two principles about eating meat: Nothing is rotten and don't spice it. His recipe for coon, for example, is to put it in the oven and roast it. That's it. "Beaver cooks up the same way. You put the hindquarters and back, in one piece, into the baking pan and cook it at three hundred and fifty degrees for two to three hours."

Patricia takes another approach, one much too cosmetic for her husband. "My favorite recipe for beaver," she says, "is to marinate the meat in wine, garlic, thyme, and bay leaf and serve it with wild rice and sautéed mushrooms with anise seeds. The nice thing about beaver roast is its consistency. It's unlike muskrat, which is stringy and awful and yucky. But beaver does smell pretty bad when you cook it."

Anderson adds that, "Of course, the other gourmet food you get out of beaver is beaver-tail soup. You have to boil the tail first." Patricia winces noticeably.

Anderson has a reputation as an omnivore. "My kids know that I'll eat anything." And, from time to time, they have tested him. Once, one of his children took him to a restaurant in New York's Chinatown. During the day, Anderson had been looking at city pigeon, meat he has eaten and likes well enough.

That night, the Chinese meal was esoteric and came in three courses.

"I didn't like the soup too well," Anderson says. "The next course was slices of sea cucumber served up with dried belly of pike. That was bad enough, but not the worst. Then came a bowl of steaming stuff with vegetables. I scooped some on my plate—oh, God, I get sick even thinking about it—there were duck's feet in it!

"The duck's feet were probably the only thing in my life I tried to eat and simply could not." They were the texture of nose. "Oh my God, weren't those things terrible!"

Patricia, having fun now, says, "I still don't know why you had such a problem with them. I thought they were inoffensive."

The stream of buyers into L.L. Bean is undiminished, though the light on the huge store is fading. One question is lurking. It remains to be addressed about a certain species. A small inquiry puts the subject on the table.

Patricia says she has been thinking of recipes for it. Her husband has a more defensive attitude.

"I really don't want to mention it," Anderson says. Finally, however, he does, and there is a pregnant silence.

"You can't print that," he says.

Suffice it to say that it is acutely embarrassing for a past commissioner of Maine's conservation department to crave, absolutely crave, the flesh of a species he has been very active in preserving.

# LOVE
# FOR A
# SKUNK

IT WAS IN THE SPRING of 1970 and Lucy Martin was driving along the "Airline," a fabled roller coaster of a road that crosses the virtual wilderness between Calais and Bangor. She was headed to Saint John, New Brunswick, to catch a late train to Halifax, Nova Scotia, where she was a graduate student at Dalhousie University.

She wasn't feeling very good about herself. Even though she had obtained a scholarship to pursue an advanced degree in literature, she felt guilty about going to school. Her family in Portland was in financial distress. She had given up her job at an insurance company to go back to school. That meant she had also given up contributing a sizeable part of each week's paycheck to her struggling parents.

Lucy was in a blind phase of her young life. She had no idea what she was going to do with her education. She felt confused, even despondent, and vulnerable. She hadn't said a word in an hour to her friend, Paul, who was driving the old VW bug.

They were about five miles out of Calais and slowly negotiating one of the Airline's treacherous curves when Lucy spotted an animal in the road.

"It was a skunk and immediately I knew something was wrong," she says, recalling the episode. "Its head was too white. Its head was encased in something white and it was tapping along on the pavement. This grotesque image came to mind of a blind man with a white-tipped cane feeling his way across a city street."

Lucy, who was raised with many animals and has great empathy with them, screamed at her friend to stop. She had an overpowering feeling that she had to do something to help the animal.

Paul obliged and pulled over about twenty yards behind the skunk, which was still sightlessly feeling its way along the rural highway. "What can you possibly do to help that skunk?" Paul asked. "He'll spray you if you get anywhere near him."

But she was firm. "We just can't drive away and leave this animal here. Its head is stuck in something. It can't see. It will suffocate. How can it eat with that thing on?"

"Forget it, Lucy, you're nuts," Paul said and started to drive away. But she became very adamant.

"Stop the car, damn it! If you're worried about my getting sprayed and stinking up your car, I'll take all my clothes off before I go to it."

Ripping her coat off her tall, slim frame, she opened the door of the moving car. Paul groaned loudly and stopped.

She got out and yelled at him that he could take off if he wanted to, but she was going to help the skunk, which by now had groped its way to the side of the road.

She approached the small animal slowly. She felt her stomach tighten. She knew that being sprayed by a skunk at close range is a devastating experience. She would rather face a bear.

Now she could hear it. The awful tapping noise on the pavement again reminded her of a blind man.

She sensed that she had no choice. She continued walking toward the skunk very slowly and started talking to it in the calmest tone she could muster. She had read and believed that animals responded to the tone of a human voice.

"I'm going to help you. Don't be afraid. It's all right. I'm sure we can get this thing off you," she kept repeating as she moved closer. She was talking to herself as much as to the skunk, creating a kind of spell.

"I'm sure the skunk knew I was following him. He probably heard me. He finally turned around and faced me. He immediately got in a threatening position, stamping his front feet and raising his tail."

By this time, Paul had maneuvered the car so that the headlights shone on Lucy and the skunk. "I guess Paul decided to help. He always said I was stubborn and at some point it was impossible to deter me."

She was getting close. "I felt I was reading the skunk accurately. The situation was much less tense. I was able to ignore any sense of danger to myself and finally knelt down next to the skunk.

"It had Tonka written on it. It was the cement-mixing part of a toy truck. It was the size of a small softball, rounded at one end and narrow at the other. It completely covered the animal's head and he was helplessly trapped in it.

"I put the thumb and index finger of my right hand on the skunk's shoulders and slowly put my left hand on the plastic thing. I told him what I was going to do and then I told him to back up.

"At first, nothing happened. Then he started backing up, actually following my directions. I pulled slightly on the Tonka part and it didn't give. I thought, 'This isn't going to work. It's on too tight.' I was afraid to yank the thing off because I didn't want to get sprayed.

"But I kept at it slowly and the object gave slightly. I got it over one ear and then over the other, and finally pulled it off. That, of course, was the critical moment. Once he actually saw me, would he see me as a potential enemy? Being sprayed at that range would have been hideous!

"Lots of questions came to my mind at that moment. What was the skunk actually feeling? He looked at me and I looked at him and there was a moment I don't want to make too much of. We both gradually backed away. He got a certain distance and turned around and slipped into the grass . . . his tail down."

Paul was standing beside the car when Lucy returned. He hugged and congratulated her. She felt lighter inside during the remainder of the trip to Halifax.

LUCY MARTIN'S ENCOUNTER with a skunk on the Airline seventeen years ago has become part of her life. "An incident like that doesn't occur in a vacuum," she says.

"My father grew up on a farm and never forbade us to bring in ducks and chickens and dogs and cats or anything we could catch wild. The animals actually lived with us. I feel I've always had an empathy and sympathy for animals—an ability to communicate with them—that's related to my childhood experiences."

She had been prepared to help the skunk by her

background and by her own sense of vulnerability at that moment.

But another animal encounter she had years later came to a much different end. By 1979, she was in the midst of a successful career as a writer for the *Maine Times*, the statewide weekly newspaper.

"We found a tiny baby mouse in the wood box at the newspaper office," she says. "It was almost winter and the mouse was practically dead. Everybody in the office was laughing at me because the mouse was so obviously on its last legs. I took it upon myself to save the poor, tiny thing.

"I took care of it for a day or two and found a way to give it some milk with an eye dropper. I put it in a shoe box and went back to work. It was hard, but I tried to do both—to work and to take care of the mouse.

"The morning of the third day it was cold and I had the woodstove going in my office and I got very absorbed in my work. I totally forgot about this little being I was supposed to be nurturing.

"The phone rang. It was a call from downstairs. I had to go down to check some copy before it went to press. I shoved my chair back, jumped up, and headed across the room.

"There was a sickening crunch. I knew immediately what had happened. The mouse had gotten out of the box and onto the floor and I had just come down on it and broken every bone in its body. I looked down at it. I had flattened it into a bloody pulp.

"You see, my mind was not on the mouse. It was on business. So later, when I came to have my own children, I remembered the mouse. I can't work and nurture at the same time. My nerves won't let me. If I try, somebody is going to get hurt."

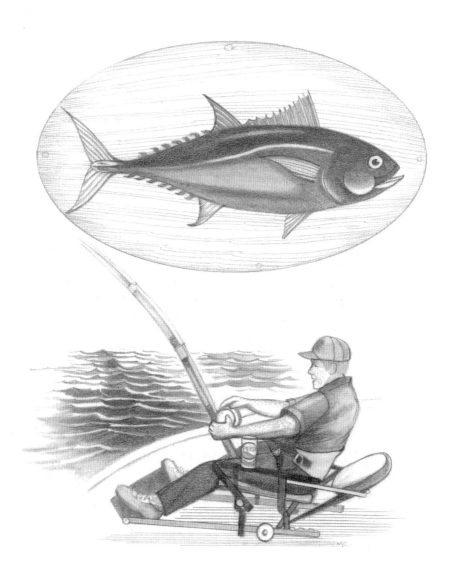

# No Lamb

SOMEWHERE THERE IS a taxidermist who loves Derril Lamb.

Among the forty or so stuffed displays in his awe-inspiring trophy room on the shores of Merrymeeting Bay are a wolf from British Columbia, a mountain lion from Colorado, a caribou from Quebec, a black bear from Maine, and a leopard, a cape buffalo, a rhinoceros, and a lion, all from South Africa.

All the magnificent beasts in the room, each a choice example of the species, died at the hands of Derril Lamb. He will not place a trophy in his trophy room unless he killed it himself with a bow and arrow.

There is a corner in the room that is conspicuously empty. It will be filled soon, though. Last May, in British Columbia, Lamb shot, with a seventy-five-pound bow, the largest grizzly bear ever taken with a bow and arrow. The thousand-pound animal's skull measured more than twenty-six inches across. When placed in its corner, the top of the grizzly's head will stand nine feet above the floor.

Lamb kills in the most intimate way possible short of doing it with his bare hands. In hunting mammals, he chooses the bow over the gun. In hunting tuna fish, one of his favorite sports, he will use the "iron," the harpoon, before he resorts to the rod and reel.

Lamb, sixty-three years old and the owner of a lumber yard and mill in Brunswick, has been hunting since he was a five-year-old. He said he started hunting "as soon as I could get a gun out of the cellar window."

Since then he has hunted on three continents. While he almost always goes bow hunting, he will take up a gun in an interlude. He had just returned from a trap-shooting championship in Scarborough. He shot seven hundred rounds in three days. He had a pressure bandage around his right wrist. The arm hurt but Lamb talked through the pain, barely acknowledging it.

Derril Lamb is a self-made man whose views are as cropped and squared away as his neatly trimmed flattop of red hair that is just beginning to show signs of gray.

"I pay my own way," he says. "I work and I've earned my money. I'm not caught up in any damned payments. I own every damned thing you're looking at. If I don't have the money to do something, I don't do it."

In 1958, Derril Lamb did not have the money to buy an oceangoing boat. So he built one from scratch, all the way from the hammering of the hull through to the wiring of the electronics.

"As soon as I got my first boat in the water, I started chasing tuna fish. I ironed four the first year and eight the second. After that I settled down into catching serious numbers of bluefin tuna every year."

When Lamb needs a harpoon, he makes one. He makes the poles out of ash. He experimented with many lengths and diameters over the years, but has settled on a pole exactly twelve-and-a-half-feet long and one-and-a-quarter inches in diameter.

Lamb, a five-foot, ten-inch hundred-and-ninety-five-pounder, throws them with precision and deadly force from his perch on the bowsprit. "I try to iron them on the back third near the spinal column. They die pretty fast."

Lamb estimates he has harpooned a couple of hundred tuna, at least. And, like all true hunters, he has a special feeling for the animals he kills. "I think they are a beautiful animal. I never saw one I couldn't admire for its shape and color."

Although he prefers the harpoon, Lamb's most memorable encounter with the tuna, which range in size from seventy-five to eight hundred pounds, was not with a harpoon but with a rod and reel.

"I'd been harpooning for years but finally decided to give the hook a try because everybody down at the dock (at Mackerel Cove in Bailey's Island) kept telling me how tough it was to catch one with rod and reel. They'd say, 'Derril, it's easy to harpoon them but you'd have a real problem trying to hook one.'"

The people who told Lamb that were "sports" from away who arrived at the annual Bailey Island Tuna Tournament with their expensive boats.

"To my mind, it's easier to catch a tuna with a hook than with a spear. You can fish anytime with a hook. But with spearfishing you need ideal weather and good visibility. You've got to be pretty sharp on the water to see those fish. They leave a little wake. It's a hard thing to see and to follow up and to get a good shot. Ideal conditions are water temperatures about sixty to sixty-five degrees with the high water slack around noon."

Those were just about the conditions Lamb had on the first day of a Bailey Island Tuna Tournament in the late sixties. But he had left his spears behind and rigged his boat, a thirty-six-foot sport fisherman named Lu-Jane, for the hook. That meant installing an elaborate "fighting chair," an item that on the expensive "sports" boats runs about fifteen hundred dollars. Derril Lamb, being Derril Lamb, built his own. The late July day was perfect. The air was hot and the water ten miles south of Seguin Island was glassy. There were schools of tuna jumping everywhere.

Lamb had his stout six-foot fiberglass rod rigged with a 14-0 Penn reel and one-hundred-and-thirty-pound-test line. He had plenty of diet Pepsi, and a tin of Bufferin in his shirt pocket. The Bufferin warded off cramps which Lamb was prone to getting after sustained physical exercise. He had little food. He liked to hunt hungry.

Lamb, as the hook specialists had advised him, had squid for bait. He was as ready as he would ever be.

The fish hit at high noon. He remembers the time exactly because it was the first tuna he had ever hooked and because his mate called in the time to tournament headquarters back at the cove. There was a prize for the first fish hooked and landed.

"It was a high point in my life," he says about seeing that huge and graceful fish break the ocean surface with the hook in its mouth. He had what appeared to be a monster on the line, several hundred pounds at least.

The fight was fun for the first hour or so. Lamb had all his strength and was enjoying the new experience of working in his fighting chair, harnessed to his expensive, heavy-duty rod and reel.

Just before the first hour was up, the tuna made a strong run and the gimbel on the fighting chair, the device that holds the butt of the rod and thus takes some of the strain off the fisherman, broke. That left Lamb's thighs as the new gimbel. Soon, they began to hurt.

Two hours after Lamb had hooked the tuna, his thighs were turning to pulp. Blood soaked the thick khaki material of his pants. His arms turned to stone. The fish was draining his strength. He dared not take his hands off the rod to sip one of his diet drinks, but thirst was becoming monumental. His partner started pouring cold seawater over his head at ten- to fifteen-minute intervals.

"I was being cooked out of my senses in the full sun," Lamb says. "I had never had such a big fish on the line before and I didn't know what I was fighting. My partner said we should cut the line, that the effort wasn't worth it. I said, 'No way, I'll stay here all night if I have to.'"

It didn't take all night. It did take, however, exactly four hours and fifty-one minutes. Lamb recalls the time exactly because his partner had to call it in, too, and because, when they finally gaffed the tuna and winched it on board, it signalled the end of an excruciating ordeal.

He was spent. He was not to be able to lift a coffee cup for three days. The inside of his mangled legs seared with pain. He couldn't even turn to look at the magnificent fish— the fruit of his immense labor—now dead in the boat. He remembered his wife once trying to describe the labor of childbirth to him. If that's anything near what I just went through, he thought to himself, it truly is the hell she described.

Lamb's tuna, at six hundred and fifty-one pounds, was to win the Bailey Island Tuna Tournament prize that year for the biggest fish caught in the rod and reel category—the first time he had ever used a rod and reel on the species.

But that night, when Lamb walked over the hill at

Mackerel Cove, he didn't give a damn about any of that. He threw away the rod and reel. He would come back to the tuna tournament, that was for sure. But he'd be returning with his spears. The tuna had ruined his fighting chair anyway, and it would be impossible to fix.

LAST YEAR, AFTER LAMB had returned from his annual trip to the Safari Club conference held in Las Vegas, he came home inspired to take a second safari. He had four of the "big five" that big game hunters want to take home from Africa, and he had them all by the bow . . . the lion, the leopard, the water buffalo, and the rhino. He lacked only the elephant.

He had tasted elephant on his first trip to Africa. The biltong had whetted his appetite. Biltong is elephant meat that is cut into strips, put into a salt solution, and set out to dry in the African shade. It was time to bag his own biltong.

Lamb started working out on his target range with a ninety-two-pound bow. Pulling such a bow is somewhat akin to breaking a softwood board along its grain with bare hands. It is possible but few can do it.

Each night, he'd shoot twenty-five to thirty arrows to get in shape for the elephant in South Africa. One night, he was drawing his twentieth arrow, when he felt a terrible pain in his elbow. He heard the muscle and tendon rip from the bone.

The arrow that would have killed Derril Lamb's elephant would have penetrated the pine and plywood of his trophy room. He doubts he'll ever shoot it now. "I think my chances of taking an elephant with a bow are over. My elbow couldn't stand it. Besides, I'm on the tail end of things. I've had too many birthdays."

He was prepared to spend $25,000 for the chance to bag his elephant. That was just for the hunt. Add to that the $1,500 for airfare, trophy fees, tips, and getting the trophy home.

"At the risk of sounding wrong," Lamb says, "I'd suggest that it's a good thing it's so expensive. The animals of the world wouldn't stand up to being taken three or four to a person."

It is a wonder they stand up to Derril Lamb.

# TROUT FEVER

"I STARTED TO FISH when I was six or seven years of age. I used to spend two weeks every summer with my aunt and uncle at Red Beach, nine miles below Calais toward Eastport. There were good brooks there, good trout. I got the fever."

The fever for trout fishing that Alden Macdonald contracted as a boy never subsided. It has burned for a lifetime. Well into his seventies, Macdonald, or "Mac" as his fishing buddies call him, is one of Maine's premier trout fishermen.

By his own conservative reckoning, he has caught more than fifty thousand fish in his life, the vast majority of them trout. He has taken thousands of fishing trips to cold-water lakes and streams all over eastern North America. The number of flies he has tied is staggering. He used to tie "a lot" of them. Now, he's down to three or four hundred a year. He passes them out to his fishing friends like chewing gum.

The mind-boggling number of fish Macdonald has caught during his lifetime is a figure sometimes questioned by other fishermen who cannot fathom the singularity of this man's passion for his sport. But Macdonald is a retired accountant. His stock in trade has always been numbers. He can add it all up for you.

"I have caught more than one thousand fish this summer alone," he said in the summer of 1987. "Of those, I have kept four." One, a six-pound, twelve-ounce trout, just brought back from Labrador, his favorite fishing spot, was on the menu for lunch.

ALDEN MACDONALD WAS BORN in Calais in Washington County in 1911. His family moved to Waterville before he finished high school. He brought with him an appetite for fishing and quickly sought out people who could satisfy it. He found older friends who quite literally regulated the flow of water.

One man, a neighbor named Bryant Hopkins, was the river engineer for the Kennebec Pulp and Log Driving Association and the Central Maine Power Company. Hopkins managed the watershed on the Kennebec for log driving and the generation of electricity. For this important business, he kept camps at the dams at Brassua, Moosehead's East Outlet, and Moxie Lake.

Hopkins took the young fishing enthusiast under his rod. He weaned him from the worm to the dry fly, the fishing artist's bait. He took him along on trips to his camps— each a site of fabulous trout fishing.

Macdonald never passed up a chance to go fishing with his mentor. And when Hopkins was busy regulating water here or there, the teenage angler turned to Hopkins's assistant in the water flowage business, Sturgis Durgin.

"Sturgis was dam tender. He regulated the flow of water from Moosehead Lake into the East Branch. He liked his whiskey, and every Friday night I'd come up with a bottle of liquor. He'd have a couple of belts and say, 'Now, Mac, I'm going out to set the gates on the dam. At four in the morning, you'll have the best salmon fishing you've ever had. I'll get those gates set so those salmon will work their way right in there.'

"He'd get up at three and pour a full water glass of that rot gut. Then we'd go fishing off the old log dam, and over the years I caught dozens of salmon.

"He showed me all the country around The Forks. He could somehow navigate in the woods when he could hardly see. We'd go six or seven miles into the woods at night with just a lantern. Back then, the bogs he showed me involved five to ten-mile walks into them. If there was no raft on the pond, we would build one. It was nothing to catch one hundred fifty to two hundred trout in one day."

Meeting and cultivating men like Hopkins and Durgin intensified Macdonald's love for trout fishing. From 1928 to the time he was married ten years later, he went fishing nearly every weekend from May 1 to September 30.

"I think I hardly ever missed," he said. And the unbelievable numbers of fish started adding up.

For the uninitiated, fly fishing, especially pursued with the verve and dedication of an Alden Macdonald, presents something of a mystery, particularly because the vast majority of fish caught are thrown back. Where is the sport in all of this?

In fact, fly fishing is considerably less sport than art. The pastime is pursued generally by people who are meticulous and exacting, by people who are immensely challenged by creating imitations of the insect world and presenting them on minimal equipment to an extremely elusive form of wildlife. It is a highly ritualistic celebration of means, not ends.

Macdonald is not overly analytical about his fixation with trout fishing. He does not ruin his avocation by dissecting it. But, when pressed, he reduces the appeal to a single image, one that makes all the travelling, all the expense, and all the winter hours of painstaking fly tying worth it: the image of a gleaming arc of trout straining for the fly.

"I like to see the fish come to the fly, whether he takes it with a smash or comes up and sucks it in. It's a challenge, really. The trout gets the fly in its mouth, and you need to set the hook before he realizes it's artificial."

Macdonald said he has had many opportunities to fish in Alaska, but he has never gone because he understands that in Alaska they take their fish beneath the surface of the water on sinking lines.

"That's not what I like to do," he said. "I'm rabid on the idea of seeing the trout take the fly."

In a corner of Macdonald's general passion for trout fishing resides a specific obsession. He has never caught an eight-pound brook trout. He won't rest until he does. (The biggest trout he has caught to date is seven pounds. He caught a six-pound, twelve-ounce trout in Ellis Pond in the Belgrade chain in June of 1932. As far as he knows, he still holds the record for that lake.)

"I've caught so many trout around seven pounds," he said. "I just want to catch an eight-pounder."

This urge, stated so simply, creates constant pressure in Macdonald to be on or in the water. He will go almost anywhere, do almost anything, to be fishing. Yearly, he and a select group of his fishing buddies journey to Labrador to Holy Grail lakes such as Osprey and Minipi. There the fish are huge. There the eight-pounders lurk.

The Labradorian trout are wild, not the product of hatcheries. Fishermen like Macdonald know the difference and expend large sums and much energy to experience it. The wild trout have an intensity of fight and spirit and a color to their flesh that hatchery trout lack.

Usually, when Macdonald fishes in Maine, he does so at a large preserve leased by a private fishing club to which he was admitted in 1976. Membership is composed of men who over the years have learned that strict regulation is the only way to preserve good fishing. Consequently, only barbless hooks are permitted. "Catch and release" fishing is encouraged, and a one- or two-fish limit is the rule.

Macdonald estimates that more than half of the one thousand fish he took during the summer of '87 came from this club in the northern Maine wilds. He is not by nature a snob. The fishing simply is good there. They let him in. So he goes.

Alden Macdonald doesn't even talk about "the fish that got away" unless the fish that was lost is, in his mind, eight pounds or better. When he talks of them, his eyes brighten.

Once he was in Labrador, fishing off a point on Osprey Lake, where the wild trout are. He had with him two grasshopper flies of different sizes.

"The water was calm and I cast the smaller grasshopper out there and the biggest trout I ever saw in my life knocked the fly four feet in the air, grabbed it when it came down, took off, and is still going! It snapped the four-pound leader on a boulder lying just under the surface.

"The next day, I went out to the same place. I had taken

the second grasshopper, the larger one, and trimmed its legs and wings down. I didn't see this one. It hit underwater so hard it snapped the leader again. It, too, is still going. I'm sure I would have met my goal, if I'd landed either one of them."

Another time, Macdonald was fishing with his wife on Lake Aguinere in northeast Quebec. He was fishing an outlet where the water was fast and shallow. His wife was fishing nearby on the lake's edge.

"On my first cast, the line hung up on the bottom. I tried to pull it loose but couldn't. I knew I had to break the leader and I was about to do that when the rod started to throb and the line started moving.

"I screamed at Margaret to bring the net. I hollered that I had a fish better than eight pounds at the big eddy in the outlet. Before she got there with the net, I put a little pressure on the fly. The fish cartwheeled twice and spit the hook."

Why, the uninitiated asks, if you want to catch an eight-pound fish, don't you use heavier test line and flies whose hooks can stand the escaping frenzy of a huge, wild brook trout? Macdonald blinks across a gulf of incomprehension. If the man cannot catch an eight-pound brook trout on four-pound line, using ultra-delicate flies, he will not catch one. The victory would be meaningless.

MACDONALD WAS EATING his trout for lunch, and the eating of the only fish he brought back from Labrador prompted the telling of a strange encounter on his latest trip to wild trout country.

"I was on Osprey Lake and I was working a trout I figured to be about three pounds. I had it pretty well to the boat when it took off and sounded. I played it a little longer and suddenly it felt a lot heavier. Maybe it was much bigger than I thought." In fact, he had a twelve-pounder in the form of two fish!

"A nine-pound pike had the three-pound brook trout broadside in his jaws. The pike just wouldn't let go. It wasn't hooked, but refused to give up its prey. With the exception of a few teeth marks, the trout was unharmed and we released it. But we killed the pike. I just couldn't believe it; nothing like that had ever happened to me before."

HE WON'T LET YOU leave without showing it to you. You close your eyes and the weight it adds to your hand is that of a college ring. It is seven feet long and weighs exactly two and three-eighths ounces. It is a cork handle attached to a melding of bamboo and graphite. The twelve-hundred-dollar fly fishing rod is a triumph of the minimalism that is the hallmark of Macdonald's sport. He has three of them. He has already broken one.

The rod is the creation of Cecil Pierce, a Southport lobsterman, shipwright, master machinist, and maker of fly rods coveted throughout the world. Pierce, now in his eighties, is one of Macdonald's oldest fishing friends. He describes Pierce as "not a college-educated man, but smarter than most college graduates.

"Cecil can sell as many of these as he can make. But he won't sell you a rod unless you are a good fisherman who will use it and appreciate it. Just because you have twelve hundred dollars doesn't mean you can get a rod from him."

Macdonald didn't buy his rods from Pierce; he was given them. He regards it as a high honor that his friend asked him to be his field tester. One master acknowledging another.

Mac sets the rod down tenderly by the window and turns to his fly-tying bench, an orderly collection of fur and feathers and tiny hooks and the surgical equipment needed to transform these raw materials into perfect imitations of the insects to which trout will rise.

It is here that he will revise the established fly patterns, some of which, like the legendary Daisy Mae, are his original creations. Something he creates at that table will, he hopes, finally seduce the eight-pounder. And if he catches that, surely there must be an eight-and-a-half- or nine-pounder out there somewhere. That thought, the image of the flash of an even larger fish straining for his fly, will keep the fever burning.

# MAINE MAN MEETS MERMAID

WOODIE HARTMAN ACTUALLY FELL in love with a mermaid named Lavaliere. That happened shortly before he became the first zoologist to witness sea cows mating and to describe the male masturbation ritual of the Florida manatee, one of the most bizarre animals on earth.

As a field biologist and longtime guide of wildlife tours around the world, Hartman, a resident of Brunswick whose father and grandfather taught at Bowdoin College, has had many strange encounters with wildlife.

Now in his forties, Hartman barely knew what a manatee was when he graduated from Williams College in the early sixties. After enrolling in a Ph.D. program in vertebrate zoology at Cornell University, he finished his course work quickly and headed for Africa where, for his dissertation, he planned to study small mammals.

"I followed the route of Lawrence of Arabia, who was, and still is, my hero," Hartman says. He fell in love with Africa but his work with the small mammals was continually frustrated. "The Africans kept stealing my traps, so I lost interest."

He returned to the United States in 1967, enchanted by the tropics but with no dissertation. He went to his advisor at Cornell. He thought he wanted to study sea turtles, but the expert in that field had his graduate students and no time to take on more. His advisor suggested the Florida manatee, an animal no scientist had ever studied underwater.

"If you can find a place to study the manatee underwater," the advisor advised, "then by all means do it."

Hartman immediately went to Florida on a reconnaisance mission. He knew the challenge would be to find a place with clear water with manatees in it. One of the most frustrating things about manatees is that they live mostly in turbid waters, which render them all but invisible.

The young scientist also learned that there was very little documented about the Florida manatee, so named to distinguish it from its relatives that live only in tropical waters. The Florida manatee has a subtropical range from the Carribbean to the Carolinas. It survives in water with a broad range of salinity but is very sensitive to cold water, which can induce hypothermia. In winter, the Florida manatee seeks out warm artesian springs.

The manatee, it is thought, evolved from an amphibious creature that grazed in marshes many eons ago. One branch of the species moved to land and became the elephant. The other branch moved to water and became the manatee.

The animal is, in Hartman's words, "a cross between a cow and a blimp." It has a pancake-shaped tail and thick, elephantine skin. It has a "hideous but fabulous" face with huge lip pads with bristles on them, which these herbivores use to tuck vegetation into their mouths. They have regenerating molars, unique among mammals, and they are huge, weighing one thousand to fifteen hundred pounds.

When Hartman arrived in Florida, his first stop was Crystal River on the Gulf Coast. He went to a dive shop which was adorned with large pictures of manatees. Hartman talked with the shop owner and checked out the springs at the headwaters of the Crystal River. He toured the rest of Florida but found no place with more potential.

So he went back home, got married, crossed his fingers, went back down to Crystal River to set up shop, and waited. He waited quite a while.

"Finally, one day on the water, I saw some noses. I was in a little aluminum punt looking over the gunwale through a glass-bottomed bucket. Through the bucket loomed my first manatees.

"They were huge. I was terrified. I asked myself, 'How will I study them if I'm too afraid to get in the water with them?' A few days later, I saw a couple of them swimming near the spring. I got in the water with my suit and mask and I saw these two gray forms turn around and start swimming toward me. I was petrified and leapt back into the boat. I said to myself, 'I'm really in trouble now. My research and reputation are ruined because I'm afraid of these creatures.'"

A few days later, Hartman tried again. He found a large group of manatees swimming around the headwaters of the Crystal River. "I got in the water and they immediately started swimming toward me. I just froze. They just swam disinterestedly past me. After that I was okay. As it turns out, the manatee is the gentlest creature on earth."

It was Woodie Hartman's job to learn as much about the manatees' behavior and ecology as possible. Every day he would don a dry suit (which doesn't let the water touch the skin) and float on the water with his snorkel and mask and observe the mammals "as long as my body could take it," usually about three hours.

In his year and a half on the project, Hartman logged huge amounts of data about manatees. He recorded everything he saw and earned his Ph.D. from Cornell on the basis of the dissertation he submitted. He learned enough to become the world expert on the Florida manatee.

When Jacques Cousteau wanted to do a program on the sea cow, he hired Woodie as his advisor. Hartman also published a piece on the manatee in *National Geographic,* which had given him a small research grant to help him in his work.

Documenting the sexual behavior of an animal is an important aspect of any such project, and far more interesting than, for example, recording its feeding habits.

"I made lots of discoveries about their behavior, but the most exciting had to do with sex. Two discoveries were so novel they called for champagne celebrations.

"The first time was after I witnessed copulation. The cow in heat was pursued by absolutely crazed bulls, driving her to distraction."

It seems, Hartman explains, that the female manatee gives off sexual attractants for a long period of time but is actually receptive to the bulls for a short time. She is, in vulgar but useful parlance, a kind of aquatic "prick tease."

The bulls get the scent of a cow in heat and suddenly she has a retinue. "I've seen as many as seventeen males all looking for a chance to get under one cow. They nuzzle and embrace her and drive her crazy ... sometimes to the point of stranding herself.

"This mating fervor waxes and wanes but when it

reaches a crescendo—when the female is finally receptive—is the only time I ever saw aggressive behavior in manatees. The males collide with one another, squeaking and squealing, and bounce off one another with those great rolls of 'blubber.'

"The strongest male usually emerges to inseminate first, but after that a thing rare in nature occurs. Many of the others get a chance also. I've even seen some subadult males get to copulate if they are in close attendance."

The natural result of there being so many sexual attractants in the water, but so few females willing to satisfy a male, is that manatee bulls have enormous sex drives which they often satisfy with each other.

"The second time I celebrated with champagne was when I first witnessed homosexual behavior among manatees. *National Geographic* warned me they would cut out all references to sex in the article. They did! I've often thought that ironic. The place most men of my generation saw pictures of bare breasts was in *National Geographic*."

There have always been rumors, Hartman says, that the manatee gave rise to the mermaid legend.

"You can imagine sailors, racked with scurvy, hallucinating, seeing a head pop up in the water after not having had a woman for God knows how many years," Hartman muses. "But you'd have to be a delirious, star-struck sailor to even consider consummating a relationship with a manatee!"

Hartman's manatee project really got under way when he gathered enough information about the manatees at Crystal River to name them. Most of the manatees have a pattern of scars resulting from propellor cuts—the pleasure boat being, along with cold water and flood-control gates, the major threat to the Florida manatee.

These scar patterns helped Hartman identify individuals.

He gave each manatee he could identify a name associated with something southern. He called his favorite animal Lavaliere, which is a term describing a rite peculiar to southern universities: If a young woman is lavaliered to a fraternity man, she is somewhere between going steady and being "pinned" to him. A lavaliere is a pendant on a necklace.

"Lavaliere was among the group of manatees that appeared regularly at the spring. Some of them liked being scratched and Lavaliere was the friendliest.

"We developed this routine: Every morning I would go out and I would try to find her. She would often come up to me. I think she could distinguish me from other divers, presumably through chemical cues in the water. We would have mutual "grooming" sessions of half an hour to an hour, lolling on the surface.

"She would kiss my mask and make barely audible little groans and grunts. I'd reciprocate; it was wonderful therapy. If everyone could spend an hour with Lavaliere, we would have a very fine world.

"Lavaliere would also investigate by chewing. She would chew all over my dry suit and my wrists and my fists.

"Then one day she grabbed my ponytail. She tucked it in her massive lip pads as if it was a piece of vegetation."

Lavaliere, Hartman's ponytail firmly in her possession, dived and took the young man down with her. For a few long moments he was sure he would drown. "It also crossed my mind that there might be substance to the mermaid myth!"

Hartman is a strapping fellow who once challenged the charge of a silverback gorilla by beating on his own chest and mimicking the gorilla's sounds—while urinating involuntarily. So he was able to yank himself free of Lavaliere's perhaps amorous embrace.

"We did go back to our sessions together," Hartman says. "But I always kept my ponytail away from her mouth after that.

"You can make out of the encounter what you want."

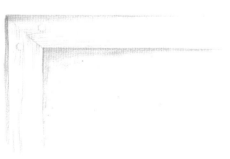

# THANKSGIVING RATS

SNOW STARTED FALLING softly at dusk on the night before Thanksgiving. Jack Aley's dog, Sirois, had been dead for a week. The dogcatcher had tossed its body on Jack's front stoop and knocked. "Here's your dog. Got hit by a pickup down the road. You're lucky I'm not fining you. There's a leash law, you know."

It took several days after Sirois died before the half-breed's strong smells began to evaporate from the house. Only then did the noises become audible to Jack. They weren't big noises, just occasional murmurings and scratchings . . . nothing louder than the gentle scraping of branches on the roof. But the sounds seemed to start deep in the cellar and slowly spread upward.

Jack had never heard noises quite like these. He was a city boy who had moved to the Maine countryside to find something hard to do . . . something physical. Rebuilding a hopelessly rundown cape in Bowdoinham seemed hard enough.

He was alone this Thanksgiving eve. His lady friend had gone back to Boston. He'd taken a long run in the snow and sweated out what was left of him in the sauna he'd built even before he had started rebuilding the house. He fixed himself a light supper of rice and stir-fried vegetables and drank a bottle of red wine. Later, he took some light reading and a heavy shot of scotch to bed. A former English major, he couldn't read heavy stuff anymore. He liked books with lots of plot now. He fell asleep before midnight.

The noise that woke him was not a murmuring or a rustle. It was a loud crash in the kitchen. Jack jumped out of bed and groped his way down the ladder from his loft. Still groggy from sleep, he stumbled into the kitchen and snapped on the light over the sink.

Two enormous rats were inches away from him, snout-deep in a bowl of Golden Delicious apples on the kitchen counter. They looked up, startled by the light. Jack stepped back reflexively. Rats occupied the same place in his subconscious as black widows and black mambas, other creatures he had never seen. These rodents looked nothing like the cute little mice his father caught from time to time in the liquor cabinet of his suburban Chicago home. The rats had tails like small ropes.

His fear quickly combined with rage. Quite suddenly and simply he wanted to kill these intruders. Barefoot and dressed only in a nightshirt, he ran into the shed and grabbed the first sharp thing he could find . . . a ski pole.

When he returned to the kitchen, the rats had disappeared. Nonetheless, Jack went wild. He started jabbing the ski pole blindly and repeatedly in machine-gun coverage of the kitchen wall behind the counter. There was a large ceramic plate leaning against the wall near the sink. It shattered at the first blow of the pole.

They came at him like two missiles of fur launched from horizontal silos in the wall. Jack jumped back, all six-foot-five of him pivoting about a foot into the air. As he jumped, the ski pole came down with the blind force of a defensive reflex. When he landed, he saw two pools of blood at once . . . under his foot and under one of the rats.

"I got the bastard!" he screamed. The buddy of the now-twitching rat had scurried away under the kitchen counter.

What to do with fresh rat kill? He didn't know right off. He prodded the little beast a couple of times with the pole and stared at it in fascination. Then, in a daze, he went to the shed and brought back a shovel. He scooped up the rat body and headed out back to the edge of his woods. Using the shovel as a catapult, he launched the rat as far as he could into the blackness.

It was still snowing and there was about three inches on the ground. He figured the rat would be covered soon, and that somehow reassured him. He walked back into the house, slowly becoming aware that his bare feet were frozen.

He was shaking from cold and shock when he got back inside. He started cleaning up the blood. He hoped it would get his mind off the rats and how he'd reacted. God, he'd gone crazy. He could feel the rat's blood as it saturated the sponge. It was still warm. He threw the sponge away when he was done and poured boiling water and bleach on the spot.

He had removed the spot from the floor, but the image of the rat twitching to death was etched into his mind.

The thought that his house was filled with rats infested his brain. He couldn't go back to sleep. He began to ask himself how he'd gotten that big gash on his right big toe. Had he merely torn the skin during his pirouette of fear and revulsion? Or had the rat bitten him on its race to impalement? Once Jack entertained the notion that the rat had bitten him, he couldn't let it go. The thought churned in his head and the thinking led in one direction: What if the damned rat was rabid?

He fell into a fitful sleep just before dawn and awoke exhausted about seven a.m. Normally he fixed a big breakfast with ritualistic care. Thanksgiving morning, he forced down a cup of coffee and called the emergency room at the hospital.

"I may have been bitten by a rat last night," Jack said, mustering as calm a tone as he could. "I'm worried about rabies. What should I do?"

The doctor, young by the sound of his voice, confessed that he didn't know if rats carried rabies. "But to be on the safe side you'd better keep the rat and have it examined at the state lab as soon as possible."

It had been a really dumb move to throw the rat into the woods. Jack got dressed warmly and headed outside with a rake. It was still snowing lightly and the day was a somber gray—bad light for the task ahead.

There now was five or six inches of snow on the ground. Jack had launched the rat into a section of woods entangled with brush and slash from his weekend woodcutting. He walked over the area carefully, hoping the rat had snagged on something and would be visible above the snow. It was not, so he started raking.

He scoured the area for three hours. His rake constantly caught on the mineral and vegetal, but nothing animal. He took a break about noon and called the homesteading couple down the road who had invited him for Thanksgiving dinner.

"Hi, David. I'm sorry I can't make it this afternoon. It

sounds weird, I know, but I may have been bitten by a rat last night and like a jerk I threw its body into the woods after I killed it. I called a doctor in town and he said I should find the damn thing and have it tested for rabies.

"All I've thought about all morning is those shots they give you in your stomach. They're supposed to be the most painful shots imaginable . . . three-inch needles right through the wall of your stomach. I don't have any appetite anyway. Thanks anyhow."

He had some more coffee and warmed himself by the woodstove for a few minutes, then headed back to the woods. It was pushing one p.m. and it would be dark in a few hours. He knew he wouldn't sleep that night if he didn't find the animal.

Jack figured the rat couldn't be more than fifty or sixty feet from where he'd slung it into the woods. With the rake, he'd gone over all the likely spots at least once. Before he started again, he tried to recreate what he had done last night. He got out the shovel and launched a couple of small plastic bags, each filled with about a pound of sand.

They landed in nearly the same spot, about thirty feet over the wall and near the base of an old apple tree. He focussed his new efforts there. Using the tree as a hub, he sifted spoke after spoke of snow and forest debris.

He was in a near-trance when the garden rake finally hit flesh. He felt the difference through the wooden handle. The tines yielded slightly, held a moment, and the rat popped out of the snow, falling beside his right foot, the one with the gash in it. The gray rat was sleek and fat and its flesh was still unfrozen. Its expression—its death mask— almost seemed a smile. Jack lifted the rat into a coffee can he'd brought with him into the woods. He left the can on a workbench in the shed as he trudged back into the house.

He poured himself a scotch and slowly chewed two cloves of raw garlic. He often did this after a workout in the winter. He sat down next to the stove, which slowly was eating its way through a late-fall diet of poplar. He was tired. The elation he felt at finding the rat was fleeting. How long did it take for signs of rabies to show?

He supposed that question would be answered tomorrow when he drove the coffee can to Augusta. Meanwhile, he needed sleep. He nodded off on the floor next to the stove and slept the night through . . . oblivious to any small sounds surrounding him.

His visit to the lab in Augusta the next morning was brief. The technician said the results of the test would be ready next week, but in the meantime Jack shouldn't worry too much. Rats in Maine, as far as he knew, had no history of rabies. Still, the lab worker warned that rats carried a multitude of other diseases and, if Jack's house was infested, he'd better get rid of them.

Jack figured if he needed to exterminate the rats, he'd better learn a little about them. On the way home he stopped by the state library to do some research. There was plenty of material about rats, and what he learned seemed to justify his fear of them. Rats were among the most tenacious survivors on earth. They were extremely aggressive and virtually omnivorous. They ate their own kind. They were programmed to destroy. Their teeth grew continually and they had to chew constantly to keep them pared. They swarmed and infested and duplicated themselves at astounding rates. There was, Jack learned, a rule of thumb about rats: If you saw one, there were at least six more out of sight.

So, he calculated, he had at least twelve rats to deal with at home. But he instinctively felt there were many more than that. And they were, at this very moment, breeding in his house.

It was late afternoon on the day after Thanksgiving when he arrived home. He'd made several stops along the way to pick up materials for his extermination project. He would spread the lime and crushed glass along the walls of the basement. Rats didn't see well and tended to feel their way along with their bodies. They stuck close to walls. Lime and glass hurt their feet. He also bought a half-dozen rat traps and a heap of rat poison. If all that didn't work, he vowed to get a cat. And he detested cats.

He parked his old Volvo and walked toward the house. It was snowing again, softly. He hesitated before opening the front door. His heart was pounding and he'd broken into a light sweat. As he opened the door, he thought he saw a blur of motion across the kitchen floor. He closed the door behind him and pressed his back against it. He tried to compose himself by taking long, deep breaths.

Then he listened. He couldn't hear them at first over the sound of his heart. But slowly the telltale rustling and murmuring seemed to envelop him. It was as if he had entered their home, not his.

He stood perfectly still for about five minutes. Then he reopened the door and walked back to his car. He drove south, to Boston. It would be an unannounced visit to his lady friend. He would return in a few days when it was sunny and he could let some cold, cleansing air into his place.

And he would bring her with him if she would come. He didn't want to be in his home alone.

# MOST EFFICIENT KILLER

WHEN THE DIRECTOR of the Brookfield Zoo in Chicago once wanted to photograph live fishers in the wild, he naturally called the Maine Inland Fisheries and Game Department. The folks there naturally told him to call Norman Gray.

Norman Gray, a tall, well-built man in his mid-seventies and a native of Fryeburg, is a lifelong trapper and a font of information about fur-bearing animals in the Maine woods. He has trapped all the native fur-bearers including several thousand beaver, "more coon than beaver," nearly a hundred bobcat, and several hundred fisher. At the age of seventy-five, Gray trapped three fisher and received almost a thousand dollars for the pelts. That same winter, Gray and his son took forty-seven beaver through the ice.

His exploits in the Maine woods are the stuff of legend: During World War II, he ran a prisoner-of-war camp at Seboomook at the head of Moosehead Lake. After the great fires that ravaged southern Maine in the late 1940s, he helped arrange for eighteen sawmills to process the burned timber. He cruised the timber on the right of way for the first section of the Maine Turnpike; that means he walked the fifty miles from Kittery to Portland. Before Mount Katahdin became the crown jewel of Baxter State Park, Gray cruised timber on the mountain for the Great Northern Paper Company, his first employer out of the University of Maine forestry school.

Wherever he went in Maine and whatever he was doing, Norman Gray managed to find time to trap and hunt. In January of 1936, he was working for Great Northern near the headwaters of the Saint John River. He found that a bobcat had been feeding on a deer carcass. He set a trap for it.

He trapped the bobcat and shot it in the head with his .38 revolver. He had the twenty-six-pound animal between his knees to spring the jaws of the foot-hold trap when the animal came to life. The bullet had merely grazed the side of the bobcat's head and knocked him out.

"He came alive. I had the axe with me and I had to dispatch the bobcat with it, holding him between my knees," Gray says in his uninflected voice. "It was a little different experience."

Back in the forties, Gray was supervising the driving of pulpwood and logs down the East Branch of the Penobscot. He had to walk six miles to get to the river. One day he was hiking through a big burn in the woods and came between a large sow bear and her cubs. The sow charged him and Gray dropped the bear with his revolver when she was thirty feet away.

He brings out a necklace he made for his wife with one bear tooth on it. The tooth is exactly one and three-quarter inches long and it came from the sow near the East Branch.

"When you see a mouthful of these coming at you, you do something about it," he says.

Gray also proudly shows a necklace made of beaver teeth. Over his head, hanging from a rafter in his camp, is the skin of a huge bobcat. His wife and daughter have coats of muskrat. His wife also has an otter hat. Gray himself wards off the winter cold with a beaver coat. When he climbs out of bed in the morning, he digs his toes into beaver pelts that serve as rugs. At Thanksgiving, the Grays feast on mincemeat made from beaver.

So when the man from the Brookfield Zoo in Chicago was directed to Norman Gray to help him photograph live fisher, he found a highly experienced outdoorsman whose passion for trapping and hunting touched every detail of his life.

The zoo director was Gregory Baker, a Chicago lawyer whose avocation was the photography of wildlife. He had photographed virtually all the major species of mammals in North America except the fisher.

There are reasons the fisher, or black cat, proved so elusive to Mr. Baker. A member of the weasel family, it is a solitary, efficient, and aggressive killer. The fisher is remarkable for its dazzling speed, sharp teeth, and sharp, unsheathed claws.

Not surprisingly, the fisher is at the top of the food chain. John Hunt, dean of wildlife biologists in the state of Maine, an expert on fur-bearing animals, a student of the fisher for almost fifty years, is in awe of these creatures.

"Nothing preys on them in the wild," Hunt says. "Nothing I know of can handle the fisher. A coyote [an animal several times larger] would lose out to a fisher." But Hunt explains that because the fisher is not afraid of anything, it is very vulnerable to trapping. It has no fear of traps and will eat almost anything in one. The only animal higher on the food chain is man.

The fisher's luxuriant dark fur has long fetched high prices on the pelt market. During the Depression, Norman Gray could make two to three months' pay with one good female fisher pelt. One he trapped last winter brought three hundred and forty dollars. The fur is used for stoles and for decorative trim such as collars.

The quality of the fisher fur coupled with the animal's fearlessness toward traps conspires against it. It disappeared from many parts of the country altogether and, according to John Hunt, was all but gone from Maine by the mid-1930s. So the state in the late thirties closed the trapping season on fisher. And it remained closed for two decades. Then, without man to prey on it, the fisher made a very good comeback.

Norman Gray was a major force in the decision to make fisher legal to trap again. By the middle 1950s, Gray says fisher had gotten so thick in his part of western Maine that "you couldn't set up a trap for a fox or a coon because you were sure to find a fisher there. We had to give up trapping for other stuff."

Gray started freezing the fisher he found in his traps and taking them to Augusta to show the members of the legislature how abundant the animal had become. Then he helped write the legislation that opened the season on fisher again. It passed without opposition.

Today, it is the fisher population that determines the length of the trapping season. "The history of the fisher's near-extirpation made us wary," says biologist Hunt. "Now the trapping rules we go by for all fur-bearing animals are based upon what we know the fisher can stand."

This policy all but guaranteed there would be fisher for Gregory Baker to photograph when he arrived at Norman Gray's home in January of 1966. Gray had prepared for the visit. He had prebaited camera sites with beaver carcasses all around the densely wooded towns of Fryeburg, Stowe, Lovell, and Sweden.

Baker and Gray took the cameras to the woods and set them up for night photography of fisher feeding on the traps. But the nights were extremely cold and the shutters worked too slowly to get good pictures.

Baker returned to Chicago disappointed in his equipment but ecstatic that he had met someone who could lead him to fisher. He revamped his cameras and returned the following year. "He stayed a few days, but it was late spring and the crows and squirrels and chipmunks and everything were out, so inevitably something tripped his camera's shutter before a fisher got to it."

The third year, Baker got his pictures. He was so grateful that he invited Gray and his family for a week's all-expenses-paid visit to the Brookfield Zoo. "The second night we were there," Gray recalls, "the curator gave us the keys to the whole zoo. We had the run of the place."

Soon after his return to Maine, the Brookfield people called Gray again. They had decided to add fisher to their collection, and would he please trap nine of them live and ship them to Chicago? They would offer him three hundred dollars per fisher.

Gray was an old hand at taking fisher, but catching them alive presented a new challenge. The Havahart traps he first tried were just no match for the fisher. So he designed a fifty-pound box lined with metal and fitted with an oak door. That's what was necessary to keep in a fisher.

It took three winters of trapping to get all those fisher

out to Chicago. The one he remembers best is one he got and then let go.

"We were trapping off a dirt road in April. It was a rainy and stormy night and it was a long hike into the trap. My partner and I were both working days, so we had to go in at night. There was a fisher in the trap, and we lugged it back out to the road—at least two miles in the dark.

"We drove six miles to my house and took the fisher into my trap house. We'd caught a female that obviously had just given birth. Her teats were swollen and there were signs of lactation. We knew she had to be taken back or her young would die."

So Gray and his partner took the mother fisher, worth three hundred dollars in captivity, back to where they had trapped her in the rainy, dark Maine woods. "I think I got back home at one in the morning," Gray remembers.

Gray says that each time they caught one of the fisher they'd hold it a few days to make sure it was healthy and then take it in a special cage to the airport in Portland. "The people at the airport weren't very happy to see us. They'd heard stories about the viciousness of the fisher. I think they'd have rather seen a bear."

Gray has noted over the years that fisher "will take food whenever it's available" even after they've just been trapped. When he was trapping the fisher live for the zoo, he'd pass whole muskrat legs into the cage and watch them disappear instantly. "If it had been my finger, that would have been eaten too."

Once Gray saw a fisher start to work on a human finger. "It was when trapping them was illegal and we got a fisher in our foot-hold trap when we were trapping other species. We handled it like a fox.

"With a fox, you grab hold of the hind legs while the

front foot is in the trap and kind of stretch him out. You can do that with a fox. You can wait for the right moment with him stretched out like that and grab for the neck, restraining the animal. That way you don't hurt the fur.

"Well, a fox is quick, too, so my partner went for the fisher like he would the fox. He waited for a good chance and went for the neck. But the fisher was so quick he got my partner's thumb in his mouth."

This was a small fisher, about four pounds (grown males run up to fifteen pounds), and he bit through the thumbnail and through the flesh and into the bone. Gray's partner danced in agony while the little fisher fastened his teeth deeper into his thumb.

He yelled for help and Gray had to maneuver with his trapping partner until he could get a hit at the fisher's wedge-like snout with a stick.

"Yep, he bit right down through the bone and all," Gray says. "Those fisher teeth are fine and sharp."

Their claws are formidable too. A few years ago a young man from Bethel was found dead under a tree with claw marks all over his back and shoulders. There was a theory that fisher attacked the man and he died of fright.

If the fisher has a well-deserved reputation for vicious efficiency in killing, it has become a legend for what it can do to porcupines, that miracle of passive defense.

No other wild animal is known to be a routine killer of the big, quilled member of the rodent family. In fact, the White Mountain National Forest in New Hampshire once tried to transplant Maine fisher because the park had a porcupine problem.

The fisher attacks the head and face of the porcupine while constantly circling. It jabs and feints, wounding and finally exhausting its prey. The process can take up to thirty minutes. Then the fisher goes for the unquilled stomach and rips it open with its teeth. It consumes the heart, lungs, and liver in its first feeding frenzy, and gets to less delicate regions later. When Norman Gray finds a porcupine body in the woods, all that is left is the skin.

It is with this background on the fisher that John Hunt's most memorable encounter with a fisher has special meaning.

"It was about ten years ago, just when the fisher was really beginning to spread around the entire state again. A bunch of us biologists were surveying deer yards up north of Moosehead.

"We needed airplanes to get around up there because there were no roads. The department had hired a bush pilot, Bob Bacon, to get us around. He was a good bush pilot, but he called all the shots and he was very meticulous about his plane.

"I'd spent the day with Sidney Howe while Francis Dunn and Hank Carson worked together. We'd been out all day and were sitting around in camp after dinner. For some reason, Francis started talking about fishers. Francis said it was really no big trick to track a fisher down and actually catch it by hand.

"Sidney began to laugh at him and call him full of bull. They went back and forth for a while and Francis finally said, 'You want to make a bet that I can't catch a fisher like that?' I remember it was a sizeable bet, more money than I would ever have bet, but Sidney finally agreed to it.

"Then Francis gets out of his chair, goes over to his pack basket, takes out his jacket and unwraps it, and there's a live fisher all tied up.

"It seems Francis and Hank found fisher tracks that day and followed the tracks to a hollow log and flushed the fisher right into Francis's coat that they held over the opening in the log."

There was then, John Hunt relates, a curious hush that fell over the campsite. The eyes of the wildlife biologists fell on the face of a very conservative and utterly appalled bush pilot.

"You should have seen the look on Bob Bacon's face," Hunt says, "when he realized that he had been flying around the northern Maine wilderness all afternoon with a live fisher in the cockpit of his plane."

# SEXING THE DUCKS

MORE THAN FORTY YEARS is a long time not to let somebody in on a joke, but for all that time John Hunt has wondered exactly what happened on a cold November night in the 1940s in Lincoln County. He has never asked, however, and he never intends to ask.

For that same period, Ransom Kelley has known *precisely* what happened that evening in Wiscasset after a hugely successful day of duck hunting. But he has never told John Hunt, a lifelong friend, nor does he intend to do so in person.

"Why should I?" asks Kelley. "Why spoil a good story?"

Ransom Kelley, in his middle seventies, is a lifelong duck-hunting guide and farmer with a furrowed and somewhat savaged face. He is a man who needs to have the upper hand and usually manages to get it. You sense quickly that he does not suffer fools.

John Hunt, in his late sixties, is a research wildlife

biologist for the Maine Inland Fisheries and Wildlife Department. He has a calm, slightly professorial manner, and you have the sense that he'd gladly suffer a fool if he had a chance to educate him.

Fate threw these very different men together early. John Hunt's father owned a share in a duck-hunting club in Massachusetts to which Ransom Kelley's father also belonged. When the Depression hit, the Hunts moved up to Maine to help build and farm for the Kelleys.

"I was standing next to John Hunt's grandfather when I shot my first goose in Duxbury, Massachusetts," Ransom Kelley recalls. "I've known John Hunt since he wore three-cornered pants. John and his father came up to work for me in 1934. When John was old enough to go to the University of Maine at Orono, I made sure the finances were available to send him up there."

So Hunt and Kelley had already known each other a long time before the "Sexing of the Black Ducks" affair, the good story Ransom Kelley didn't want to spoil.

This incident might be more telling given a couple of examples of Ransom Kelley's brand of humor. The first story is what Kelley did to a friend, the second to an "obnoxious sport."

"I called up John Baxter one night and said I needed some help the next morning shooting a few geese. Baxter was a relative of former Maine governor Percival Baxter and was on the Fish and Game Advisory Committee for years—he was a very hungry hunter, if I do say so.

"So we went out the next morning early and were set in the goose blind in the river, and I told him the geese would be along between 7:05 and 7:10 a.m. At precisely 7:07, we heard the first squawking.

"I saw the first geese out in the distance and gave a few

toots with my tooter, and they came up and swung into the wind over my decoys. There were five of them. When they came around, I told John to start shooting on the downwind end and I'd start on the other and we'd meet in the middle.

"We got all of them, and then I told him to get out of there quick because we had exceeded the two-bird limit. He was hurriedly loading up the back of his car when he asked the question: 'Ran, how did you know those geese would be along exactly when you said they'd be?'

"And I said, 'For Christsake, John, how does a good guide know anything?' For the next two years, every time I saw him, he'd go back to that morning and ask how in hell I knew exactly when those geese were coming.

"What I never told him—for at least five years—was that the night before I had been hunting and a flock of six geese came, and I intentionally shot the gander out of the flock. I knew enough about geese to know that the females would be back to that very spot the next morning."

If there was a better-known duck- and goose-hunting guide in the Merrymeeting Bay area over the last half-century than Ransom Kelley, it was his father-in-law, Earl Brown. One day a number of years ago, Kelley and Brown had fixed to go gunning on the coast. But an older couple had shown up and Brown's wife had promised them that her husband and son-in-law would be happy to take them hunting.

"That made us a little unhappy," said Kelley, "but we took them. We were setting up down by the water and I had forty decoys draped over my shoulder when the male sport comes up to me and says, 'Hey, my good man, lug my lunch box for me.' I wanted to tell him where he could shove his lunch box, but I didn't."

Earl Brown and Ransom Kelley set the man and

woman up to shoot and then had a brief conference.

"Every time a duck came near and the old man got ready to shoot, Earl brought his red handkerchief to his nose. After a while, the old fellow had to go into the woods. Earl stopped "blowin' his nose" and the wife gets three quick ducks and her husband comes back on the run.

"Earl immediately goes back to blowing his nose with that red handkerchief and the old guy doesn't get a shot and goes back into the woods again to get warmer. We did this four times. The wife got her limit of ducks and some for him. He didn't get any. He died that following winter.

"Two years later, I told the woman about what we'd done with the handkerchief to drive off the ducks. She got the biggest kick out of it."

HERE'S HOW TO SEX a black duck. Pick it up and lay it on your knee, and take your thumbs and place them on either side of the duck's anus, then squeeze the anus with the thumbs. If a small organ a little thicker than the lead in a pencil pops out, you've got a male. If nothing happens, you've got a female.

The reason you need to sex black ducks or "snap their penises," as some guides like to refer to the process, is that it is all but impossible to otherwise distinguish between female and male black ducks.

Sexing ducks for John Hunt in the middle 1940s was part of his job. He would check with hunting parties at the end of the day and inspect their bag for sex and age. The information would be used to help formulate waterfowl hunting policies.

John Hunt, by his own admission, took himself very seriously back then. He was a member of the third class ever to graduate with degrees in wildlife biology from the University of Maine.

On the November evening in question, Ransom Kelley was in the back of the Ledges Inn in Wiscasset tending to his ducks when John Hunt showed up to do his job. Kelley had had three drinks and eaten his supper and was basking in the afterglow of a great day of duck hunting. His six sports had each gotten their limit of ten ducks and had done so early in the day. There had been time to relax. There had also been time to sex the ducks.

"How did it go, Ran?" Hunt asked, walking up to his older friend.

"Oh, we had awful good luck today, John," replied Kelley. "And John, no need of you to go snapping the penises on these ducks. All we shot was males today because they were so plentiful."

Well, Hunt gave his canny friend a knowing look and went over to the pile of ducks and performed the telling ritual on every one of the thirty or forty ducks there. Every duck was a male.

Hunt finally stood up and scratched his head and asked his friend how they could tell the males from the females in the air.

"Oh, that's easy," replied Kelley, "the females go 'quack, quack, quack' and the males go 'hiss'."

Kelley remembers, "He kept looking at me trying to figure out what the kicker was. But he never asked me outright. It wouldn't have done him any good if he had."

John Hunt knew something wasn't right. "They had me buffaloed. To this day I don't know what happened to the females. But at this time I was trying to be scientific and I remember writing up some kind of report."

And what happened to the report John Hunt started to

write up about the sex and age distribution of a pile of male ducks? He never filed it.

What had happened to the females, of course, is that Ransom Kelley, after sexing the ducks, sent them home with the sports who left first, thus setting the stage for the trick.

John Hunt, when he reads this, will find out for the first time how he got tricked.

# DOES A BEAR...?

THE CALL CAME on a late November day in 1944.
"Hello, Mr. Leavitt. My name is George Stirnweiss and I'm calling from New York. I'd very much like to go deer hunting in Maine, and a friend of mine said you might be able to help set me up."

Bud Leavitt, now a well-known Maine outdoors writer, was then sports writer for the *Bangor Daily News,* and he just about swallowed the phone.

"THE George Stirnweiss?" he blurted.

"If you mean the baseball player, yes, Mr. Leavitt."

THE George Stirnweiss happened to be the star second baseman for the New York Yankees. That year, he had won the team batting title wth a .319 average, and was also baseball's leading base stealer.

For a Maine sports writer to get a chance to go deer hunting with George "Snuffy" Stirnweiss was like a parish priest having an opportunity to say mass with the Pope. Leavitt was excited.

"Fine, fine, Mr. Stirnweiss," he stammered. "You want to go next Monday? Good. I'll set it up for you. You'll enjoy it. Just come into the paper and ask for me. I'll see you then."

Before Leavitt told his colleagues who his deer-hunting buddy was going to be next week, he placed a call to two fish and game wardens, Moses Jackson of Bradley and Earl Peasley of Orono. These wardens had taken Leavitt under their wing when he had started writing about the outdoors, and the three men had become friends.

"They more or less nurtured me when I started," Leavitt says. "I didn't know anything about the outdoors. I was just a sports writer.

"I told them that the New York Yankees' star second baseman was coming up to hunt and asked them if they could help set something up. Well, they didn't know who Snuffy Stirnweiss was and cared less."

The wardens had access to a camp called Crocker Turn, deep in the woods of Greenfield, which was about forty-five miles north of Bangor. They said Leavitt and the baseball player ("Snuffy what's-his-name") were welcome to use the camp. They'd meet the pair on Monday afternoon to get them set up for hunting on Tuesday.

Over forty years later, in retrospect, Leavitt says it probably would have been wiser not to tell the two wardens that he was bringing a famous baseball player up with him.

Wardens Peasley and Jackson were a kind of a team. They went after poachers together and patrolled together, and occasionally stalked and killed troublesome wildlife together.

It so happened that just about the time George "Snuffy" Stirnweiss was beginning his long drive to Maine to meet Bud Leavitt, wardens Peasley and Jackson were tracking down a bear that had been giving fits to a sheep farmer in their territory. They shot and killed the marauding black bear, a medium-size male, late Sunday afternoon.

Then the wardens called a butcher friend who had a freezer. They told the butcher that one of the wildlife biologists at the university had asked the Fish and Game Department for a freshly killed and frozen bear. The biologist wanted it for research.

The butcher watched with some fascination as Moses Jackson and Earl Peasley wrestled the beast into his deep freeze. They didn't just throw the bear in there. Peasley carefully propped the bear into a sitting position. And Jackson used some tape to keep the bear's eyes wide open. The butcher assumed that was the way the biologist wanted the bear to be delivered.

As the wardens were leaving the freezer, they looked back to inspect their handiwork. The bear, sitting in the corner and just beginning to stiffen, looked weirdly human. Its dead, dark eyes, so closely set in its keglike head, stared at them. Peasley shuddered as he closed the heavy freezer door.


LEAVITT SAID HE COULD TELL it was Snuffy Stirnweiss coming in the newsroom because he'd seen the star's picture come across the Associated Press wires so many times. Leavitt introduced Stirnweiss to his fellow reporters. Then, it being midafternoon, the sports writer and the ballplayer left directly for the camp.

Stirnweiss started to unwind the minute he hit the rustic enclave in the woods of Greenfield. He told Leavitt he had been dying to get away for weeks. All the postseason publicity he'd gotten in the wake of his batting championship had worn him out. Too many long interviews and too many bad chicken dinners. He was looking forward to a week of hunting in the woods with no distractions and no crowds.

Stirnweiss met Moses Jackson and Earl Peasley as soon as he arrived. They seemed kind of cool to him, but the celebrated baseball player found that refreshing after all the fawning attention that had been paid to him.

Leavitt had told Stirnweiss that Peasley had a reputation as a great camp cook, and Snuffy was starved after the long pull from New York. He stowed his stuff in the cabin, lighted a predinner cigar (a nice Havana), and strolled toward the kitchen. Something smelled awfully good, so he popped in to ask Peasley what he was preparing. What Leavitt had not warned the second baseman about was that Peasley "made a lot of noise about people invading his 'inner sanctum'" when he was cooking.

"You're off base, Stirnweiss. I don't allow anybody in this damned kitchen when I'm cooking," Peasley roared in response to Stirnweiss's appearance. "I don't care if you're a New York Yankee baseball star or who you are. Get out."

Poor Snuffy couldn't figure out what in the world he had done to so provoke the warden. "He was absolutely crushed during the dinner hour," Leavitt recalled. "He didn't say a word."

After the meal, Stirnweiss retreated to a corner of the rough dining room and relit the cigar he'd started before dinner. He quietly watched the wardens as they prepared to go out for the night in pursuit of poachers. Peasley had not referred to the incident in the kitchen, and both wardens seemed studiously to be ignoring their famous guest. The ballplayer sensed tension in the air and was on edge. He'd never been treated like this before. He was beginning to think these guys were giving him a hard time just because he was a big-name athlete.

In his preoccupation with all this, Stirnweiss had been ignoring a basic urge. Suddenly he had to go and go badly. But where? Reluctant to approach the inhospitable wardens, he went up to Leavitt.

"Say, Bud, where's the bathroom around here anyway?"

It dawned upon Leavitt that his guest, a native of New Jersey, probably had never used a two-holer in the woods before. So he took him outside as far as the woodshed, gave him a flashlight, and pointed the way to the outhouse, which was about a hundred yards away.

Stirnweiss thanked him and scurried across the distance as fast as his short, muscular legs could take him.

The ball player probably sensed immediately upon entering the dark outhouse that he was not alone. But he was in no condition to exchange amenities. He went straight for the empty seat, dropped his brand-new hunting pants, and sighed.

It was a warm night for late November in Maine—in the upper forties—and Stirnweiss probably felt a strange chill. There was also a foreign smell that had nothing to do with outhouses. Stirnweiss couldn't place it.

"Kinda raw in here, isn't it?" he remarked to the silent presence on his left. "I wonder how the hunting will be tomorrow. I've come a long way for it." But there was no response and, not for the first time that evening, Snuffy Stirnweiss began to feel uneasy.

Curious about his neighbor, Stirnweiss snapped on the flashlight and directed its beam to the other side of the two-holer.

The thawing bear's head was inclined perfectly to pick

up the full shaft of the beam. Snuffy found himself staring into the most malevolent gaze he had ever seen.

If Joe McCarthy, then the Yankee manager, had witnessed the next ten seconds of his second baseman's life, he'd have had heart failure. Wardens Jackson and Peasley later swore it was the loudest scream they had ever heard in their lives.

Stirnweiss, his new hunting pants at his knees, crossed the distance between the outhouse and the cabin in far less time than it took him to steal second base.

"There's a b-b-b-b-b-bear in the toilet," he croaked as he crashed through the cabin door. Leavitt says he had never seen a man so white with fear: "He was literally sweating."

Peasley looked up slowly at the athlete and commented: "Where else would you expect a bear to go to the bathroom in the middle of the night in the Maine woods?" Peasley then started a slow, quirky smile, one that was duplicated across the room on the face of Moses Jackson. Leavitt had not been let in on the secret but quickly added it up; he'd been victimized by his friends before. Then Leavitt started to smile.

"What the hell are you smiling about?" Stirnweiss cried. "There's a b-b-b-b-bear . . . " He stopped midsentence and looked slowly from one mischievous smile to another in the small, smoke-filled room.

"Oh my God," he groaned. "I've been had."

George "Snuffy" Stirnweiss had indeed been had. He had also been initiated, and he took it so well the wardens warmed up to him quickly. They guided him to the best deer hunting in the area the next day, and Snuffy got a nice doe within hours. They even let him go out on their poaching detail one night.

"Whatever those two wardens told Stirnweiss to do,

he'd jump to do it," Leavitt recalled. "He really fell for them." It seemed that the ball player was hugely impressed by the scope and execution of the elaborate joke that had been perpetrated on him.

He was so impressed that he just had to tell a few of his friends when he got back to New York. How could you keep a story that good to yourself?

JOE MCCARTHY ALWAYS HAD a few words to say to his boys at the beginning of spring training every year in Saint Petersburg, Florida. In the spring of 1945, he saved what he had to say for Snuffy Stirnweiss, who was poised to have another good year. Indeed, he was to lead the American League in batting (.309), hits, stolen bases, runs scored, and triples.

According to Leavitt, McCarthy tore into his second baseman for behaving so irresponsibly during the off season. The manager was fuming. "All the rest of the guys on this team worked hard all winter getting in shape and taking care of themselves and you, Stirnweiss, risk everything running away from a dead bear in the Maine woods in the dead of night.

"Did you think of the consequences to this team if you'd broken an ankle. I ought to fine you." McCarthy then glowered in silence as the eyes of all his teammates fastened on Stirnweiss. The young athlete felt much as he did the night of the incident in Maine—he wished he could disappear. But then the smile that slowly spread across McCarthy's face reminded him of the smile that had spread across the face of Maine warden Earl Peasley.

About two weeks later, the Yankees were playing the Chicago White Sox in an exhibition game in Saint Petersburg. It was the bottom of the third and Stirnweiss had ripped a single into left field.

On the third pitch to the next batter, he attempted to steal second and was called out. He jumped up and bellowed to the umpire: "You're blind, ump! I was safe!"

The umpire paused long enough to get his face right up to Stirnweiss and then he bellowed back loud enough for all the players on the field to hear: "Any dummy who can't see the difference between a live bear and a dead one ain't in no position to challenge one of my calls!"

They say that Snuffy Stirnweiss didn't win a challenge all season and that, after a while, he stopped trying.

# DAMN MUSSELS

"DAMN THE COLD. They could freeze, you know."

Mickey Varian, the trucker, spoke. Dain Allen, the fisherman, nodded. "Yeah. Been bad. It was a bitch getting them."

They were leaning against the truck watching the sea smoke swirl around Allen's wharf in Harpswell. Varian, six feet tall and swarthy, paused to catch his breath next to his shorter, stockier friend. They had just loaded more than two hundred bags of mussels, each weighing sixty pounds, into Varian's ten-wheel Diamond Reo. It was late afternoon on a January Sunday. The temperature was near zero and plunging.

Mickey Varian of Sebasco was poised to make his third round trip in seven days to Fulton Market, one of the world's largest fish distribution centers, which is located in New York City, an eight-hour haul from Dain Allen's wharf.

Allen and two of his men had worked for two days to

harvest enough mussels to fill Varian's truck. In the intense cold they'd been dredging mussel beds on a Casco Bay island just off Freeport. Allen was as exhausted by his repetitive work with steel dredges and chain bags as Varian was by his hard marriage to the road.

There was irony in how the common mussel, a humble bivalve, drove these men. There was little market for it in New England, where historically it had been used as bait. But there was a profitable gourmet market in New York and elsewhere around the country. The trick was getting them there time and time again.

In some ways mussels are hardier than men. Intertidal animals, they are tough survivors in the harshest environment in the sea. They endure great swings in wetness, temperature, salinity, and the friction caused by waves. They beat these odds in part because they are hugely prolific. Their genitals dominate their anatomy. It has been estimated that an acre of flourishing mussel beds will produce up to ten thousand pounds of mussel meat per year. (An acre of good pastureland might produce one to two hundred pounds of beef). So Dain Allen would always have plenty of mussels to harvest and Mickey Varian plenty of mussels to haul.

Varian hadn't always been a trucker of seafood. Once, he'd been a full-time fisherman like his old friend Allen. But he had gotten tired seeing so much of the money from the fish he caught going to the middlemen. It was a struggle to support his wife and seven children. So he became a middleman himself.

He did most of his own driving, not just to save on expenses, but so he could tend his accounts in person. An independent, he had no union and drove himself in a way no union would permit. At thirty-nine, he was still relatively young but there were times that he felt old.

Dain Allen was getting over a recent divorce, so he had decided to accompany his friend Mickey on this trip to New York, a city he hadn't seen in twenty years. He wanted to "get back into circulation" again.

Varian and Allen pulled out of Harpswell about four in the afternoon. The interstate still had ice on it from two recent storms, but Varian kept the hammer down as much as possible. He was in a race with the cold. He didn't know how long the Coleman heater in back with the mussels would work.

The old friends talked of trucks and fishing and women and fights and the goddamned weather, and their talk was punctuated by frequent and unintelligible bursts from the Cobra CB.

Varian spotted Buddy French's rig at a truck stop in Massachusetts. French greeted them with a voice like a gravel road. He'd been in the bar all day recovering from a trip to Alabama he'd begun only four days before. He'd really been rolling, he beamed, and he had more than three hundred dollars in cash on him. He told Varian and Allen that they were going to help him drink up that money. They did, for three rounds, until Varian pleaded with French for understanding. "I got mussels freezing in the back of my truck."

Back on the road for the long last lap into New York in the worsening cold, the talk failed between the men. Allen tried to grab some sleep, but the frost on the inside of the window kept burning his cheek. He got restive. "Hey, Mick, when are we going to start seeing all those high buildings?"

The Diamond Reo crept into the frozen city through the Bronx about three in the morning. Varian made a slight detour to show Allen the prostitutes, but the cold had driven the ladies indoors.

The world-famous Fulton Fish Market was locked in one of the coldest nights in New York City history. The smoke hugged the surface of the East River, and a thermometer perched on a high building registered two below zero.

But even in the cold the market lived. There were trucks everywhere and from everywhere—from Orlando and Beaufort and Provincetown and Montreal. All along the two-block length of the market, fires roared in steel drums. The scores of Italian workers who unloaded the trucks stoked the fires with crating wood. The men stomped around the blazes in tight circles before rolling their snow-clogged carts back to another truck.

Under the large, open shelter and along the street were littered signs of the market's commodity. In one gutter there were two stiff sturgeon, a giant slab of halibut, and a small flounder.

Varian parked his rig and sought out Nunzie, the portly boss of the group of tough-looking Italians who always unloaded Varian's truck and got thirty cents a bag for it.

Mickey had gotten to know a number of men in the market, and they waved and smiled when they saw him in his red-and-green tam and rubber fishing boots.

The Maine men found Nunzie huddled over a straight whiskey in Carmine's bar, one of the several old and oily and beautiful bars around Fulton Street that rose out of the moil of the market. Nunzie was glad to see Mick because he'd gotten a nice little bonus from him at Christmastime. Nunzie told him he would have a couple of unloaders right away.

Soon, Al and Charlie were stamping their feet next to

the truck and damning the cold in two languages. They were heavily swaddled and carried the obligatory steel hook over their left shoulders so the sharp point caught just below the shoulder blade. All the workers at the market carried their hooks that way.

Varian opened the back and found the trailer awash in a slush of seawater that had leaked down from the small mountain of mussels. It appeared his worst fears might be confirmed because many of the heavy bags were frozen together. It would take heavy forearm blows to dislodge them.

Varian had wrecked a truck the month before, and had hurt his shoulder, so he tagged the bags while Allen delivered the blows and hauled the bags to the hooks of Al and Charlie, whose job was to carry them from the back of truck to the buyer's storage area.

They worked for a couple of hours, taking occasional breaks to huddle by the burning barrels. Even in the cold, it was possible to work up a sweat. Nothing, however, helped their feet, which seemed to depart from their bodies.

By first light, the work at Fulton Market was complete. Varian always felt some satisfaction after delivery, but the cruel conditions of this run made its end especially sweet. At dawn, the two Maine men trudged back to the seemingly ever-open Carmine's. Varian had his usual, a blackberry brandy, and Allen downed a double whiskey, neat.

On the way out of the city, they were treated to a spectacular sunrise behind the skyline. They passed the Statue of Liberty and Varian said he never saw "the old girl" without getting a proud feeling inside. Allen nodded and then said he might come back to New York sometime but never again with Mick in the truck. There were easier ways to get back into circulation after a divorce, he was sure.

When they were three hours south of the Maine border, Mickey Varian perceived and heeded the signs of exhaustion: he twitched; every few seconds his right wrist flipped up involuntarily; periodically, his chin slumped to his chest and then started up, as if jerked by an invisible string. He couldn't even keep the accelerator down; his foot was losing all sense of feeling. He gave the wheel to Allen and fell into a fitful sleep.

They reached Harpswell in early afternoon to find that Dain Allen's step-brother, "just to show Dain up," had gone out and come back with four hundred and nine bushels of mussels.

The exhausted again encountered the inexhaustible. Mickey Varian never even got home to kiss his wife or see his seven kids. He loaded up the mussels and called another driver and asked him if he wanted to go to New York. At sunset, Varian started on another eight-hundred-mile round trip to Fulton Market with a truckload of mussels. Dain Allen, the fisherman, said he'd be going out for more in the morning.

# THE DANGEROUS SPORT

THERE WAS ONCE, the story goes, a game warden who lived way down east, and after he retired he did a lot of sea duck hunting. He had spent some time as a warden searching for lost sea duck hunters. So, every time he would go out, he would tie himself to the offshore ledge before he started to gun. He said that if he ever drowned while sea duck hunting, the search party would find him near his boat and tied to his ledge. "They'll never have to drag around and be cold and miserable because of me," he said.

Sea duck hunting is one of the most dangerous sports in Maine. A state biologist who has been hunting sea ducks for thirty years observes: "The only time I go hunting and feel that my life potentially is in danger is when I go hunting sea ducks."

The danger inherent has nothing to do with the prey. Though they are harmless creatures themselves, sea ducks—eiders mostly—live on ledges and small islands off the coast. The season for hunting them is late fall and winter when an accidental tumble into the numbing ocean water can be fatal. Add the ingredients of fog and ocean swells and giant tidal swings. This is a very risky environment.

Some people think sea duck hunters are crazy—masochistic even. Who in their right minds would intentionally expose themselves to such a harsh environment to kill marginally edible birds? One hunter tells how he and his buddies were transformed into ice men by freezing spray on the trip back from a ledge. They had to break the ice off each other after they got to shore and then pour rum down each other's throats. They cherish the memory of that day.

Gareth Anderson of South Harpswell doesn't fit this stereotype of the sea duck hunter. He's an articulate man, tall and dark with hawklike features, who makes his living being careful and teaching caution. He is the safety officer for the Maine Inland Fisheries and Wildlife Department. "I don't ever intend to be in danger," he says. "In my profession you get gun-shy in a hurry." He's the kind of person who, before every duck-hunting season, packs a change of clothing into a five-gallon plastic bucket and ties it in his skiff.

Nonetheless, Anderson is an avid sea duck hunter. The eider, especially, attracts him. It's a miracle of game management that the large sea bird is still around to do so. In 1910, by one estimate, there were only fifty pairs of eider duck left along the coast of Maine. Years of hunting and gathering their eggs and disturbing their nests for their valuable down all but wiped them out. The state closed the season on eider for many years. They came back beautifully.

The eider now is so numerous its season is longer than for other ducks. Gareth Anderson likes that. The eider represents an extension of his hunting days. He also likes the bag limit, which is seven, and the fact that one hunter can have fourteen in his possession at one time.

"The males are pretty," Anderson says. "They're white and have a long, leathery projection on their bills that runs onto the forehead. They've got a little dusting of green on the back of their heads and a buff-colored breast. They're the biggest duck around here. They run five to seven pounds.

"They're not the best table bird, so I take the skin off to get rid of the fat and get right at the dark red meat. I never met anybody who didn't like my eider stew."

Anderson tries to go eider hunting at least once a year with his son Robert, who is a fisherman in South Harpswell. In October, after the lobstering slacks off, Robert asks his dad to go out for a day. They invite a few friends and make a big stew or quahog chowder that they keep hot on the back of the boat with a Coleman heater. It's an annual outing, often to ledges in nearby Middle Bay.

Anderson remembers one such outing vividly: "We left here early on a weekend day in late October in 1984. We were on Robert's thirty-two-foot Nova Scotia. It was terrible foggy. We started jogging out through Middle Bay and Robert read his compass and started nudging the throttle up a little bit. We were looking to go out to Little Whaleboat Island, where I guess there's been duck blinds for two hundred years.

"We were going at a pretty good jog through this awful fog and I start to get a little nervous. This boat costs forty to fifty thousand and it's his livelihood. I started to say, 'Why are you hurrying? We're just going sea duck hunting. There's no hurry. And it's awful foggy.' But I didn't say anything. He spends all his time on the water. I'm not going to tell him how to run his boat. But we're going seven or eight knots and I'm still nervous.

"I'm standing by the pilot house on the starboard side looking around, and there's this damned island just towering above us. It's Shelter Island, which is one of those bold islands. It makes up just like cliffs on all sides.

"I let out a hell of a yell, but it was too late. We piled right up, high and dry. We hit a mossy incline and it jacked the boat right up. A couple of us went sprawling. The dragger sat there perfectly balanced, when by all rights it should have toppled over.

"Instantly, my dog was gone. His name is Shaman, and he's trained to go when the boat stops. So he's over the side thinking we're going duck hunting.

"I thought that at least we'd busted a keel. We had about thirty minutes to high tide. One of us got in the skiff to use as a towboat, three of us got under the bow to shove, and the dragger's propellor was still in the water. So Robert gunned it full speed in reverse. The boat didn't budge.

"We decided to rest and wait and be fresh for the tip-top of the tide. One of Robert's friends is a giant, and he got under the bow and we did it all again and got the boat off the rocks. We checked the boat for leaks and didn't find any, so we went duck hunting on Little Whaleboat.

"It turned out that all there was was a little nudge on the bottom. I'm sure what saved the day was the mossy incline and Robert's quick turn of the wheel after I yelled. Still, the incident dulled the day. We had a good shoot but talk would return to the episode every fifteen minutes or so. It's one thing to endanger your boat while working, but another when you're playing."

MANY YEARS BEFORE this incident, the Anderson men also had been sea duck hunting in a fog. With them was a friend of Gareth Anderson's, Bob Clark of Thomaston. Clark didn't like the water much. It wasn't his element.

They were south of Ragged Island off Bailey Island, and heading to Bold Dick, about six or seven miles offshore. When people on the coast say "bold" they mean vertical, and Bold Dick is a pinnacle of rock that rises straight out of the water. An absolutely inhospitable place and therefore perfect for sea ducks.

"Because of Bold Dick's configuration," Anderson remembers, "the water is always rough and choppy around it. It's real hard to get on. Whenever I hunt around it, I always keep someone in the boat so we're never unprotected.

"We were in a dungeon of a fog. Ducks would come up on the decoys fast and make a couple of quirks and keep going. All of a sudden these wraiths would appear out of the fog and be gone. We never got a single shot at them.

"I figured conditions might be better in Hell's Kitchen on the south end of Ragged Island. So we idled in there. There were lots of birds. We turned the motor off so we could listen for breakers. It's a rough, exposed spot. We shot a couple ducks and we were in the outboard so we could go pick them up.

"Suddenly, I had the sensation of something coming up through the ocean real fast. But there was nothing coming up. We were falling. We were going down on the water peeling back from the ledge. It was making a groaner.

"The foot of the outboard hit the ledge, and the motor was thrown into the boat with its propellor still going. The boat bounced off the ledge and wasn't rising quite as fast as the water. Then the groaner came in real high, topped with white froth and turning a light green color like they most always do, and hit us with a hell of an explosion.

"I had to stop the motor and Robert knew to go for the oars. Bob Clark was taken totally by surprise. The boat was half sunk and the water was cold and there was probably only six or seven waves until another hellacious groaner would hit us. We rowed to beat the band to get out of there!

"I'm sure Bob never looked forward to going sea duck hunting again after that. He really didn't care for that activity."

If you look closely at Gareth Anderson's hundred-pound black Labrador retriever Shaman, you'll notice his left eyelid droops. He owes this condition to the eider.

"Some time ago, four of us took the skiff I'd just built out to Long Ledge near the mouth of the New Meadows River. We got out to the ledge and put out the decoys. We got some eider right off and the dog had retrieved a bunch. The tide was rising, and some pretty good breakers were starting to come in, along with another flock of eider.

"We shot three out of the flock, and Shaman got one back to us and had started up a gradual ledge—almost like a beach—with the second when one of those groaners came up. Shaman was halfway back with the duck." Anderson was on higher ground so he had a perfect view of the huge green wave picking up his beloved black dog with the eider in his mouth.

For a moment, it seemed Shaman was frozen in the translucent green roll of water. The dog's eyes were wide open under the wave. He looked more quizzical than terrified.

"Shaman went by me fast with the eider in its mouth and the groaner broke and sent him careening onto the ledge where he dropped the duck. Well, he scrambled right back out there to retrieve that eider and when he got back this time I had a chance to check him out. He seemed all right."

Meanwhile, the wave had crashed over most of the ledge. It had upset the boat and ruined one party member's new gun. The five ducks Anderson had shot had been taken back to sea and were adrift. Shaman had to retrieve them again. "I don't know how many ducks he got that day because he had to get them all at least twice!" Anderson says.

The wave quite literally dampened the hunt and they quit early. Back home, Anderson checked out Shaman thoroughly for injury. "I couldn't find anything wrong with him except that his left eyelid drooped and does to this day when he's tired. I guess the pressure from the groaner did it."

Anderson has been on many rescue missions, most of them involving relatively inexperienced sea duck hunters returning from offshore.

"Many people just can't suspect what's out there—few have any conception of how cold you can get out there hunting sea ducks in three hours. You want to come home and build a fire and dance around it.

"The inexperienced hunters generally will go for light boats that are convenient to carry. But three people with all their gear are often heavier than the boat itself."

Anderson said that most drowned sea duck hunters have been caught by following seas in estuaries. These men are often motoring pell-mell for warmth when the groaners sneak up behind them and capsize their tiny craft. They die quickly in Maine's winter water.

Anderson is familiar with a very famous incident

involving three people—a middle-aged fisherman and his teenage son and nephew—who died while hunting sea ducks. He knew them personally. They lost their skiff and drowned in the high tide that inundated their ledge. Anderson has read the well-known short story about the tragedy, "The Ledge," by Bowdoin College professor Lawrence Hall.

"My wife's father bought the dead fisherman's gun," Anderson says. "They found it on the ledge."

# Twenty—One Days in the Woods

GEORGE WESCOTT SLICED the golf ball deep into the woods on the first hole of the Massachusetts golf course. When he had gone golfing before his ordeal, he didn't slice the ball so often. The loss of half of each foot made a difference in his swing.

He sighed and made his way the few yards to the golf cart, another change in his game. He used to enjoy the walk. Now he was incapable of it. There was no question of his foraging around for his drive. Besides, just being near the woods, as civilized as these woods were, made him think back several years to a morning when he had entered some very uncivilized woods in Maine. He emerged twenty-two days later a totally changed man.

IT WAS A HUNTER'S SNOW that was falling on Greenville on November 15, 1982. Tracking the wary whitetail deer would be easier than usual. Fifty-two-year-old George Wescott of

Swansea, Massachusetts, had been hunting in Maine for eighteen years. He always looked forward to there being a little snow on the ground, and was more eager than usual to get into the woods this day.

Wescott and his hunting party were in Township 8, Range 10, part of the Bowdoin College Grant near Greenville. They drove to the end of a Scott Paper Company woods road and parked. It was eight a.m.

He had hunted with these friends for years. The rules were always the same: hunt alone or in pairs in the morning and come out at noontime to regroup. That way, if a deer had been tracked or wounded, there would be time for the group to pursue it before dark. And that way, if one of the hunters in the party got lost, there would be time for the others to search for him before dark, which comes early to Greenville in mid-November.

Wescott quickly oriented himself to the road with his compass and struck off alone that morning. He was wearing his favorite gray wool pants with the red checks, a green hunting jacket, a blaze-orange hunting cap, and a Timex wristwatch he had purchased at Harris Drug Store in Greenville.

The woods were beautiful, and the new snow was freshly imprinted with signs of all kinds of wildlife, including deer. Wescott knew that the chances of actually killing a deer that day were slim. He had shot only about a dozen in all his years of hunting in Maine. But he loved the process of trying. He always imagined there were two or three of the ephemeral whitetails just out of sight—over the next creek or behind the next hill. He moved for two and a half hours in a sort of hunting trance. He was oblivious to the rough ground and intensifying snowstorm.

He was not sure how long he had lost the feeling in his

feet before he actually became aware of it. The wet snow had penetrated his boots and two pairs of socks. His hands had also become numb with cold, and he was forced to carry his rifle in the crooks of his arms.

"I've never been this cold hunting before," Wescott thought to himself. He glanced at his Timex: 10:30 a.m. "I'd better get back to the road."

He was able to retrace his steps through the snow for about half an hour. After that, the snow, which was being driven by a hard, cold wind, obliterated all signs of his tracks. He stopped periodically to plunge his freezing hands into his armpits. The pain in his hands was reassuring. As long as he could feel them, he knew he wasn't losing them.

His feet were another matter. His walking seemed almost like floating because all feeling stopped below the ankles. It was eerie to look down and see his feet move through the snow but not feel them. Eerie and frightening. The juices from George Wescott's big breakfast began to rise in his throat as bile.

Where was that damned road? He was sure he was headed right.

The next time he checked his watch it said half past three in the afternoon. Dark was coming on fast. He paused to try to stomp and rub some feeling back into his feet. When the feeling came it was in the form of searing and breathtaking pain. He looked out through the shroud of snow and saw the shore of a pond, and on it the dim outline of a camp. He moved slowly toward it. He would try to break in and spend the night. George Wescott, veteran deer-hunter, was lost in the Maine woods.

(To those who don't know the woods, it may seem strange that an experienced outdoorsman and hunter such as Wescott could lose his way. But the forest is a very tricky place. It is possible, for example, for a man who has owned his thirty-acre woodlot for ten years—and walked every inch of it many times—to get lost in a snowstorm and wander about for better than an hour before finding the bordering road. Almost every veteran outdoorsman has been "turned around" at one time or another.

Bud Leavitt, whose reporting in the *Bangor Daily News* was the basis of this story, said that he had once been lost and going in circles for a day while deer hunting. When he finally heard and followed the sound of a truck, he discovered he had spent the entire time less than two hundred yards from the road he had been seeking. The woods can swallow a person.)

By two p.m., Wescott's hunting partners had become alarmed. All but one of them went back in the woods to hunt for the hunter; that one stayed behind either constantly honking a car horn or blowing a whistle. At dark, they all assembled back at the road. There was a chainsaw in one of the vehicles. They ran the chainsaw for two full hours, hoping its piercing whine would guide their friend back to them. It didn't. They waited, and their wait became a vigil.

It was four o'clock the following morning, a Tuesday, when Wescott's friends acknowledged their need of help and alerted the Warden Service. The search began in earnest at dawn. The wardens came and swept the entire area all day. The snow had swallowed up all sign of the hunter.

Bloodhounds were flown in from the Midwest. Helicopters and airplanes were enlisted in the hunt. In the next six days, up to a hundred and fifty professionals and volunteers scoured the dense woodlands around Greenville for George Wescott.

The wardens were confident they would find the man. Man-hunting in the woods was now a science. Based on a doctor's report of Wescott's physical condition, a computer model had been made of him. The computer told the wardens just how far a man weighing more than two hundred and fifty pounds, a heavy smoker with high blood pressure, was likely to get through rough and snow-covered country. It wasn't very far, and the search was concentrated in the range the computer dictated.

Six days and twenty-one thousand dollars later, the search for George Wescott all but ceased. There was no way a man in his condition could survive at that season in the Maine woods. Thanksgiving came and with it the end of deer-hunting season, and, in the minds of the wardens and the public, the end of George Wescott. Someday, someone would run across his remains somewhere out there.

THE NIGHT GEORGE WESCOTT forced himself into the small cabin on the shore of the pond, he weighed about two hundred twenty-five pounds, forty pounds more than he would weigh in three weeks' time. He hadn't smoked much in several years and his blood pressure was okay. By his own testimony, he was a man who could stalk a deer fifteen miles in a day. He could easily outdistance the model the computer had set up.

That first night, he was not concerned about his life or being lost. He knew how good the Warden Service was at finding lost hunters. He was embarrassed that he would be the target of their search, but he didn't worry about that much either. His major concern was for his feet.

He built a fire immediately in the tiny camp stove. He had dry matches and there was a store of kindling and firewood. He thawed out a little and tried to eat a sandwich he had packed, but he wasn't hungry.

His feet were so badly swollen he had to cut off his boots with his hunting knife. He removed his stockings from his left foot and with them came raw flesh. He marvelled that the pain wasn't worse. He removed the stockings from his right foot with similar results. The front two-thirds of the soles of Wescott's feet had been deeply frozen. He gazed at his extremities in fascinated horror.

He waited in a kind of limbo for a few days. Where were they? Weren't they coming for him? Maybe he had better move. Maybe they'd see him if he tried to move. His feet were getting black, and the smell from them worsened as the feeling in them lessened. He must move while he still could.

He left the first camp and spent a night out on some pine boughs in a hell-hole under a ledge. It rained. He found a second camp on what he was later to learn was Lower Wilson Pond. He was in shopping distance of downtown Greenville. He scrounged up some peanut butter and tuna fish in the camps. A man could live, if not flourish, on that.

By this time George Wescott had been lost in the Maine woods for two weeks. He sensed that Thanksgiving had come and gone but he wasn't sure. He could be dead by Christmas. His strength was ebbing. Would they ever rescue him? He'd heard the planes a few times. He tried to keep a fire going in the cabin so they could see the smoke.

He wrote "help" on a blanket with adhesive tape and spread it outside the cabin. He hung his orange cap as high up in a tree as he could manage. He fashioned a horn out of a fishing-rod case and blew as hard as he could.

Sometimes, he toyed with the idea of how easy it would be to fall asleep on the snow and not get up. But Wescott was a survivor, and he would move again and again until it wasn't up to his will anymore.

He found another cabin on Lower Wilson Pond. There was a small sailboat there, a Sunfish. He could see what appeared to be a homestead across the lake. At night he saw a beacon which he figured must signal the Greenville airport. How could he be so close to civilization and still be so lost?

About the eighteenth day after vanishing into the Maine woods, Wescott knew his strength was all but gone and that the infection from his frostbitten feet and hands was spreading. If he didn't get across that lake soon, he was going to die.

The first morning he tried was a Sunday. It took him an eternity to wrestle the small boat into the icy lake. When he attempted to roll himself into the boat, it skidded out from under him and he fell in the water. His curse was of despair.

The next morning it dawned flat calm, and Wescott, his fingers hurting terribly, attempted to climb aboard the Sunfish again. He concentrated hugely on each move. There was no room for a slipup, and there was none.

It took him two and a half hours to paddle across the pond. He reached the dock on the other side and, unable to stand in the boat, rolled himself out. Then he struggled the short distance to a road where he saw tire tracks. He waited.

It was a telephone company truck that found him. The driver stopped and gawked at the bedraggled apparition in the road.

"Are you who I think you are?" he said to Wescott. "Are you the lost hunter from Massachusetts?"

Wescott had been lost in the Maine woods longer than any person in recent memory. It was simply unheard of to disappear in the woods for twenty-one days and survive. Wescott told his story to the Maine wardens and then to the newspapers.

He was unprepared for the controversy. The Warden Service was criticized for giving up on the search so soon and putting incorrect information on Wescott's physical condition into the computer. Most of the calumny, however, was heaped on Wescott himself. Letter after letter to the editors of area papers said that this man was a fraud. He'd done this as a stunt to get publicity for the small Massachusetts restaurant he owned. He'd done it so he could make money on a book. One trapper was skeptical that Wescott was ever lost at all: "You could smell Greenville from where he was." Others said Wescott should pay back the twenty-one thousand dollars it cost to look for him.

There was so much controversy surrounding George Wescott that the *Bangor Daily News* sent its outdoor columnist Leavitt down to Wescott's Massachusetts hospital bedside to prepare a series of stories. Leavitt got a firsthand look at his feet and a firsthand impression of his character.

"George Wescott paid a hell of a price to go deer hunting in the Maine woods," Leavitt concluded. "He's a real gamer in this book."


ON THE SECOND HOLE of the Massachusetts golf course, Wescott sliced again. He drove his cart to the edge of the woods where his ball had disappeared and reached into his trousers for another one. In the same pocket were the pictures he always carried with him.

He took the pictures out and stared at them before attempting his penalty shot to the green. The photos were of his blackened and bloody feet before he lost so much of them to amputation. He smiled slightly before he rocked back a trifle on his heels and hit a crisp five iron right onto the green.

# Pinned

MANY YEARS AGO, a man from Waterville shot a deer in the Maine woods. He had shot many deer in his life and prided himself on making the kill with one bullet. He had the habit, after dropping his prey, of leaning his gun against a tree and approaching the deer with his skinning knife drawn.
    It was late in the fall. Snow was in the air. He shot and the buck fell to the earth. He leaned his rifle against a tree and came up on the deer with his skinning knife out. He was kneeling and making his first cut, when the buck heaved in postmortem spasm. Its hoof caught the hunter in the temple. He slumped to the ground.
    They found the hunter the next spring lying in a near-embrace of his last deer kill. It had been a cold winter. Neither the body of the deer or the man from Waterville had decomposed very much.

FRITZ WAS A GIFTED DOG. The Doberman pinscher could roll a log better than any lumberman, and could retrieve anything. If there were five pairs of hunters' slippers in the camp, he would unerringly return with the pair he was told to fetch. The men in the camp loved this game and so did Fritz.

Above all, Fritz was disciplined. If told to, he would stay on the unfenced grounds all day, no matter how strong the passing temptation. And he never, ever attacked deer.

Fritz's master was Albert Thibodeau, a native of Aroostook County who by the early 1930s had been around a great deal. He was a powerful and lithe athlete who once had been part of a pyramid-building act with the Barnum and Bailey Circus. He also worked the trapeze.

During the First World War, Thibodeau's ship sank in the North Atlantic. He was one of only a few survivors. When he returned to the States, there were only two men left from his circus act. So he went home to Maine and found a job as a fire warden. He was assigned to an isolated station on Eagle Lake in the Allagash wilderness, which was fine with him. He loved the outdoors and he loved to hunt. He and his dog Fritz thrived there from early spring to late fall each year.

Thibodeau had a good friend and hunting companion, Gene Letourneau, who was a reporter based in Waterville for Maine's Guy Gannett chain of daily newspapers. The two men had a lot in common. Letourneau had also been an entertainer, a drummer who had started in the orchestra pits of silent movies and beat his way up to a big-name band in New York City.

But Letourneau's heart was not in the city. He stayed there long enough to know he had to come home. He arrived in Waterville on a Wednesday, shot a deer on Thursday, and got a job as a reporter on Friday.

His beat ranged from Augusta to the Canadian border. Within this territory was Albert Thibodeau's remote post in the Allagash. And there was no place the young Letourneau loved to go more.

In the early thirties it was an entire day's trip from Gene Letourneau's home in Waterville to Albert Thibodeau's cabin on Eagle Lake. First, Letourneau would drive from Waterville to the foot of Chesuncook Lake. Then he would take the mail boat nineteen miles up the lake to Umbazooksus Stream, where there was a tramway with a handcar powered by a Model T Ford engine. He rode that the final sixteen miles to Eagle Lake.

Thibodeau's two-story cabin was a centerpiece of the fire warden service. Letourneau felt much at home there. As soon as he arrived, he took his gear upstairs, unpacked, and placed his slippers near the other men's. The first thing he did when he came downstairs was to command Fritz to fetch his slippers. Fritz never missed. The only man closer to Fritz than Albert Thibodeau was Gene Letourneau.

It was partly because of Thibodeau that Letourneau could travel to Eagle Lake so often during hunting season. The reporter covered sports as part of his general assignment, but nobody back then devoted themselves to reporting about hunting and fishing. "Gene, you really ought to start writing about the outdoors," Thibodeau said once during an outing. "There'd be a lot of interest in that."

So Letourneau started writing an outdoors column once a week and it took hold. Now, starting into his eighties, Gene Letourneau is a widely known outdoors writer with a string of books to his credit.

LETOURNEAU WAS ON "assignment," he says, that late November day in the early 1930s when he headed up to Eagle Lake. He left early on a Friday morning. He and Thibodeau along with a few other wardens and guides in the area planned to get in a day or two of deer hunting before the season closed.

Saturday dawned gray and cold on Eagle Lake. It felt like snow. The men ate a big breakfast, assembled their gear, gave Fritz the habitual pat on the head, and scattered. They hunted alone and planned to meet back at the warden's cabin in midafternoon.

Thibodeau felt enormous confidence in the woods. He wasn't cocky about his prowess, but he knew the area well and had never made a dangerous mistake. He was a crack shot. If he saw a deer he wanted, he shot it. He never shot twice, out of pride and because ammunition was expensive. Those were lean times.

He had had a busy fall as fire warden and had spent precious little time hunting. He hadn't shot his deer for winter meat yet, and he was set on doing it that day. He knew where the best deer habitat around the lake was, bushwhacked for an hour to get to it, and began to stalk.

Thibodeau soon spotted a large buck with an imposing rack near the shore of the lake. He started to pursue it, maneuvering for a good shot. The hunter wanted a sure shot, so he wanted to get close. The footing was good. The forest, a section as yet uncut, was near its climax stage, with huge trees and little undergrowth.

Within minutes of spotting the buck, Thibodeau thought he had his shot. The deer was close, within a stone's throw. He couldn't miss. In the long instant before he pulled the trigger, the deer stiffened and turned his majestic head. There was eye contact as the gun went off.

The look in the buck's eyes was one a hunter never wants to see in the eyes of its prey. It was not surprise; it was not terror. It was somehow a look of confession. The animal was saying through its eyes, "I have made a mistake."

The bullet tore into the deer's neck slightly off the mark, but still with staggering force. The buck took a couple of steps before falling near the base of a large spruce. Thibodeau thought he had done what he usually did—made a killing shot with one bullet.

He moved up on the animal slowly. He leaned his rifle against a tree and drew his skinning knife from the sheath on his belt. In the next motion he kneeled next to the buck and started his first cut into the hide. It was his habit to pause in his first cut and wait until the fresh blood began to flow.

The buck suddenly bolted to his knees and at the same time swung his ten-point rack into Thibodeau's muscular chest. The first thrust knocked the man off balance and back against the spruce tree. Then the animal literally gathered the man in his antlers and plunged, pinning Thibodeau against the tree.

It was pure chance, of course, that the fire warden was not impaled, for he felt that this was exactly what the animal was trying to do. Instead, the rack enclosed him around the chest, its points digging into the bark of the huge tree.

Thibodeau stood five feet ten inches and weighed about a hundred and eighty pounds. He was hard as a rock and kept his muscles toned by working out on a high bar he'd rigged in his camp. The workouts reminded him of his circus days.

His struggle to free himself from this bizarre embrace

was frenzied for fifteen minutes or so. He applied all the force he could muster with the one arm that was free of the antlers to try to twist the buck's head. But the animal would not budge.

After Thibodeau had spent himself physically, he forced himself not to panic. He studied the animal almost clinically, as if it had nothing to do with him. The buck was breathing regularly, and its breath stank. The strange look in its eyes when he shot it had been replaced by one oddly far away and glazed, like the trance of an endurance athlete summoning his deepest reserves.

It was cold in the woods late in the afternoon. The snow Thibodeau had smelled earlier had started to fall. He again tried to wrestle himself free, and again failed. The buck's breathing remained regular and deep. His eyes remained open and unblinking and far away.

It was clear now to Thibodeau that the shot had not been mortal. It was clear there was no way he could free himself. And it was also clear he would not last the night, dressed as he was and with a snowstorm gathering.


BY FOUR P.M., BACK AT THE CAMP, Thibodeau's party was worried that he had not returned. He was usually the first to return from the hunt because he liked to get an early start cooking dinner.

It was Letourneau's idea to send Fritz, and he was the only man in the group with enough rapport to give the dog the order and get results. "Fritz, go find Albert! Go find Albert! Go find Albert! Now go!"

Fritz at first looked up at Letourneau quizzically. What a strange command. But he sensed the urgency in the voice of his master's friend and bounded away into the woods.

This dog could retrieve anything. Could he find Albert Thibodeau?

Fritz came upon Thibodeau in less than half an hour. The man's scent was still fresh in the woods. But the sight of his master pinned against the tree by a deer befuddled him. He was trained never to attack deer.

"Fritz," groaned Thibodeau. "Get the deer off of me. Please, get the deer off of me. You don't have to hurt him. Just roll him, Fritz. Roll him."

The back of the buck did look a bit like a log and Fritz tried to roll it, but the buck didn't roll like a log and there was no water. The buck did have ears, however, and the smart dog took one of the ears in his mouth and bit down hard and started to roll again.

In moments, Albert Thibodeau was free of the rack.

Thibodeau, leaping for his gun, had never before finished off an animal with mixed feelings. He was a hard-nosed hunter. But now there was a turmoil of emotion in him. He wished he could let this beast go free. But that was out of the question. The animal would soon die of the wound.

So he shot it.

There was no question of dragging the deer back that night. His strength was all but gone. He had difficulty enough following Fritz home through the soft fall of snow.

When he got back to camp, the first thing he did was to ask Fritz to get his slippers. Then he told his story to Gene Letourneau and his other hunting friends. They all went out the next morning to drag the buck back to camp. Its meat helped Albert Thibodeau through the winter. And its antlers got pinned on the wall.

# THE MAGNIFICENT GHOST

THE MOUNTAIN LION is the Santa Claus of Maine animals. There are believers. But the evidence we have of its existence is a tantalizing patter on the roof.

The mountain lion, also known as puma, cougar, catamount, or panther, has over the course of this century become a symbol of elusiveness and mystery, as has its far-distant cousin, the snow leopard of the Himalayas.

Long held to be virtually exterminated from the United States east of the Mississippi, the mountain lion in Maine was last "documented" just after the turn of the century. Herbert Adams, a Portland writer who has made an historical study of the mountain lion, reported in the August, 1985, issue of *Habitat: Journal of the Maine Audubon Society*, that the "last official kill" was made September 1, 1906.

Since that time, there has been no "hard" evidence— either a photograph or a body. Yet, over the years, there have been constant reports of sightings of the majestic cat.

Gene Letourneau, whose column "Sportsmen Say" has been syndicated in Maine newspapers for decades, says, "I must have received two hundred reports of mountain lion sightings over the years.

"If a guy calls me and says he's seen something, am I going to argue with him? If he thinks he saw a mountain lion, it's all right with me.

"Are they out there? I think anything is possible, especially with an animal that can travel as easily as a mountain lion. One can go a hundred miles in short order. And I've seen some awful tracks in certain kinds of snow and melt. But then the track of a ten-pound cat can look like that of a one-hundred-pound animal in the right conditions."

"Mountain lions have started to roam once again," began one of Letourneau's columns in the late 1950s. That lead or a slight variation of it has served to announce many accounts of mountain lion sightings. He has never needed to embellish to engage reader interest; that is ensured by the topic. Maine people seem enamored of mountain lion stories.

The stories have come from many places. During the 1950s, Letourneau reported sightings by a wide variety of Maine people in or near Brownfield, Skinner, Jackman, Princeton, Andover, Houlton, Dedham, The Forks, Rangeley, and Solon. In the decade that followed, reports came from Bethel, Clayton Lake, Ross Lake, Patten, and Bingham.

During the 1970s, mountain lions were reportedly seen in Somerville, Rockwood, Bingham, and Sidney; and so far in the 1980s in Waldoboro, Farmington, Telos Lake, and Temple.

That list—only partial—is extensive enough to suggest

that mountain lions possibly still live in Maine. There are, as a state wildlife biologist says, "just too many reports to discount."

Consider a sampling of the accounts reported in Letourneau's column:

In 1960, Mrs. Norman Dock of Bethel was hunting and suddenly froze in the woods: "Within a hundred and fifty feet of me was an exact replica of what I had seen in zoos and circus cages many times, a jet-black panther. I had a wonderful gun loaded with two slugs. I was an experienced hunter, but I didn't shoot. I was honestly terrified and figured if I missed, I would be in trouble.

"When I later told my husband, he spoofed, so I didn't dare mention it anymore. I often wish I would have had the courage to shoot. Then the skeptical question would have [been resolved], providing I had fired straight."

In 1976, Lewis Johnson of Sidney, a former warden with the Maine Department of Inland Fisheries and Wildlife, said he was a member of a deer-hunting party in Washington County that fired at and wounded a "strange cat-like animal." The hunters recovered some hair and flesh and blood and sent the samples to Massachusetts for testing. The results reportedly indicated they had wounded a mountain lion.

In December of 1983, Laurie Ann Vigue of Winslow reported she had spotted what appeared to be a mountain lion during the previous June near Telos Lake: "Its huge size and long tail ruled out bobcat and it was running as only a very sleek cat could run. What a sight! And I had seen it clearly too!"

Ed Hoyt of Benton reported through Letourneau in January, 1986, that he had seen one cross Route 43 near Farmington. Hoyt, who had seen mountain lions in the

Dakotas, said it was the largest he'd ever spotted, measuring about eight feet from head to the end of its tail. "I got within twenty feet of it," Hoyt said.

Later in 1986, Wilfred Wagner of Dixfield said he was driving up to his camp at Temple "when I reached a point on the road where I could see about two hundred feet ahead. At about seventy-five feet, there was an animal walking away from me. At first I thought it was a fox, then a coyote, but when it turned broadside, I knew I was looking at a mountain lion; it had a perfect profile, a small, pretty cat head with darkened ears, the sleek body with elevated hips and, most notably, the tail, which . . . curved all the way to the ground with a perfect spiral on the end. It was by far the most beautiful animal I have ever seen in the woods and I am seventy-two years old."

In the summer of 1983, an especially convincing sighting brought the Department of Inland Fisheries and Wildlife into action. As reported in the *Portland Press Herald,* Susan Hunt of Waldoboro and a neighbor watched a mountain lion for almost an hour off the Dutch Neck Road in Waldoboro. "It lay in a nearby tree, climbed down, walked around, climbed the tree again, and walked into the woods," the story said.

Since the sighting had been the third in Lincoln County in six months, the department sent a biologist and photographer to investigate the area. They found hair and ran tests that established one thing—it had been a wild cat. That, however, didn't preclude bobcat, a fairly common species in Maine which is much smaller than a mountain lion.

There are reasons why the mere prospect of mountain lions in Maine inspires interest and awe. They are impressive animals, with males weighing up to one hundred and fifty or two hundred pounds and stretching up to eight feet from head to tail. They are fast and have bone-crushing force in the swipe of paw and bite of teeth. The cats routinely kill animals several times larger than themselves. Deer, especially those found in rugged, very isolated terrain, are a favorite food.

Our ancestors did not hold the mountain lion in awe. They feared it and wished it were gone. The beast was a harsh reminder of the threatening wilderness. Some of the names early Maine settlers had for the animal, such as "Indian devil" and "mountain screamer," attest to anything but a romantic or idealized attachment.

Herbert Adams unearthed in his story for *Habitat* an account of a nineteenth-century Mainer's encounter with an "Indian devil." This was no mere sighting:

"As late as 1880 Portland's *Eastern Argus* told the breathless tale of an attack on one Granville Herrick, who tarried too long in the "Promised Land" sector of Poland, Maine. On April 14, Herrick heard wild and distant screams 'to which, however, he paid no attention,' said the paper, 'until suddenly the strange animal sprang from a tree directly on to Mr. Herrick's back.'

"Herrick, 'being a very athletic man, succeeded in striking the beast a heavy blow with an axe he fortunately had with him . . . Mr. H. turned to run, when he was seized by the foot, which in the struggle he pulled from his boot.' Bootless and bloody, Herrick roused the neighbors and returned with armed reinforcements. One promptly gave the animal 'a charge of heavy shot, which caused him to retire to the swamp with fearful yells.' Herrick's deliverer, 'a hunter of large experience, pronounced it an American Panther or "Indian Devil" . . . and thinks Mr. Herrick's escape was providential.'"

It is a remarkable evolution that contemporary residents of Maine, separated from this experience by only a century, can view wildlife such as the mountain lion so differently. To our ancestors, the "haunting wail" of the big cat would most probably have produced a shiver of fear. To most of us, that same sound, were we "lucky enough" to hear it, more likely would produce a shiver of delight and wonder.

"Perhaps there is something alluring in the thought that this majestic ghost, proud and mysterious, still walks undetected in this computerized world busy stamping out its last mysteries," observes Adams. "Of course, to skeptics this allure is misleading, and the lack of hard evidence is critical."

John Hunt, one of Maine's foremost wildlife biologists, is a skeptic who, one senses, still wants to believe the big cat prowls in some dark corner of Maine.

Hunt still remembers vividly the first time he saw a panther. It was many years ago in the Cascade Mountains of Washington State. "I found a deer kill one day. Coyotes were feeding on it and quite suddenly they fled. Shortly thereafter, a mountain lion came strolling out and picked the damned thing up in its jaws and walked off with it. Every hair on my body stood up."

It has been part of Hunt's job over the years to follow up reports of mountain lion sightings, including those reported by Gene Letourneau. Hunt says that mountain lions are "great hands at covering stuff up with their paws" and generally scratching things and making signs "that only lions make . . . In forty years, I've never seen any of those signs."

Yet Hunt says he knows at least two people in Maine "whose veracity I'd bet on and who are keen observers" who claim to have spotted the cat. "If what [they] described is what [they] saw, it was a mountain lion...

"No, I wouldn't bet a cent that there's not one out there," Hunt concludes.

# CARIBOU: THE EVOLUTION

"THE CARIBOU HAVE ALWAYS BEEN, and are now, emblematic of the North. Whenever people think of the Arctic and semi-arctic environments, the animal that most often comes to their mind's picture is the caribou, large herds of animals moving across open land.

"When you stand on the barrens of Newfoundland in the fall and see coming from the distance a mature caribou in his fall breeding regalia, with huge bronze antlers and white mane, you are inspired because it represents one of the most stunning wild animals imaginable."

Shane Mahoney, a Newfoundland wildlife biologist and a leading expert on caribou, is mesmerizing in his Irish poet's voice. He has a huge audience, the millions of people watching the *Today Show*. The eleven-minute segment, an eternity by major network standards of news coverage, was introduced by the anchorperson with the hook: "Ah, there's good news this morning. Santa's reindeer have arrived in Maine."

The *Today Show* piece was the high point of an orgy of media attention paid to the December, 1986, transfer of twenty-seven caribou from the Avalon Peninsula of Newfoundland, where they still prosper, to Maine, where they were long ago slaughtered into extinction.

This was the latest chapter in perhaps the most dramatic of all wildlife stories in Maine. The saga of the Maine caribou is a complex one of brutal, if unwitting, annihilation; subsequent guilt; and Olympian efforts at atonement. Probably no encounter with an animal better illustrates the evolution of a hunting society based on killing to survive, into a conservationist society based on protection of wildlife for its own sake.

The last small herd of caribou in Maine lived on treeless tableland on the side of Mount Katahdin in 1908. The mountain offered both the semi-arctic climate and the lichen they preferred. After the caribou vanished, they remained only in the names of mountains, bogs, narrows, lakes, and one northern Maine city.

In the fifty years preceding their disappearance from Maine, the caribou was a target of constant killing. Its meat was a favorite in logging camps all over the state, and was also marketed in Eastern cities. In the subsistence culture of the last century, the caribou was never seen as an exotic species worthy of protection, but as a food source easy to kill. There is a photo of about a dozen newly shot caribou piled in a heap on the ice of Moosehead Lake. It is reputed that local hunters surrounded the caribou and ran around them in circles until they shot every one. The date on the picture is 1903. There are also stories of whole towns turning out to kill entire herds of migrating caribou.

It has been said in our ancestors' defense that they didn't know what they were doing. The caribou moved around so much, it seemed that there were more of them than in fact there were. But it may also be surmised that had Maine hunters of the last century known they were slaughtering the caribou into local extinction, they would not have cared. The conservation ethic was many decades away.

Who knows when the guilt took hold, and in whose minds, and how it was spread? Who knows when it first crossed the mind of a Maine hunter, "My God, what did we do?"

But fifty-five years after the last caribou were spotted upon the tableland of Mount Katahdin, the guilt has been growing long enough to prompt the bearers of it to do something. The prime movers behind the caribou transplant attempt of 1963 were the spiritual children of the men who killed off the caribou. Many of them were hunters who now hunted for sport, not for survival. Professionally, they were wildlife biologists and conservationists and veterinarians. They set out to undo what had been done.

Ladd Heldenbrand was one of them. A veterinarian with a degree in zoology, he often had occasion to work with the Maine Inland Fisheries and Game Department. He knew men in the department had been trying to put together a plan to transplant caribou from Newfoundland to Maine's Mount Katahdin in Baxter State Park. He told them, "If you ever decide to do that, cut me in."

Just before Thanksgiving of 1963, Heldenbrand got a phone call at the camp near Bangor where he was hunting. "Ladd, if you want to go, we're going in the morning." The veterinarian had less than twenty-four hours to get ready for the trip of his life.

Newfoundland had agreed to exchange their caribou for Maine's wild grouse, which the province wanted to introduce to its wilds almost as much as Maine's conservationists wanted to reintroduce the caribou. The deal was ten pair of grouse for each woodland caribou. It was to take about three years to capture the grouse. The caribou were ready now, at the remote Indian Lake district of the island province.

The call came on a Thursday. "We have your animals. You've got until Sunday to come and get them or we let them go."

THE CANADIANS HAD BEEN keeping their eyes on the herd around Victoria Lake. Timing was all. When the herd started to migrate, the men had to move fast to round up a few. They had built a corral next to the lake, but the caribou skirted that and tried to swim away. A plane circled the animals in the air and then landed on its pontoons and circled them in the water. Men in boats then lassoed the caribou's antlers and dragged them to shore. It took an entire day to get about fifty caribou in the corral.

The men in Maine were not entirely prepared. "We frantically talked Coles Express into giving us two trailer trucks and two drivers," Heldenbrand remembers. Then there was a sleepless marathon of driving to North Sydney, Nova Scotia, for the ferry to Port aux Basques, Newfoundland.

There were very few roads in Newfoundland: The only way into the lake district was on a narrow-gauge railroad. "These were the largest trucks ever seen in Newfoundland," Heldenbrand says. "I couldn't believe the accuracy of the drivers backing those trucks onto the flatbeds. Their wheels were hanging over on both sides. The slightest slip and the project would have been over."

At the corral, Heldenbrand helped sedate the caribou. "We had to bulldog some of them. You stand beside them and reach under them and grab the opposite foreleg and pull them down and then you tie their legs like a dogie." The men watched in awe as one three-hundred-pound stag cleared the eight-foot fence of the corral. It had already been injected with a tranquilizer. One of the men made a note to call Maine and make sure four feet of fence was added to the height of the corral being readied in Millinocket. Reindeer could really fly.

The roads in Newfoundland were rough, often with no shoulders. The crew followed the trucks and saw their tires begin to burn. The gravel on the roads caught in the huge tires' treads and abraded the rubber until it caught fire. "The drivers changed six tires on that trip," Heldenbrand recalls.

The Royal Canadian Mounted Police stopped the trucks at three in the morning and flagged down the following car. The Mountie stuck his face in the window and said, "We knew you were coming." The Mainers groaned inwardly.

"What have we done wrong now?"

The Mounties had prepared a party for them. They'd set up a buffet in the next town, with sandwiches and cakes and lots of coffee. "We know what you are doing," they said. "We want to help."

The dockmaster at Port aux Basques wanted no part of caribou leaving Newfoundland for Maine. He had not been advised they were coming and he simply refused to let them on the ferry. The Maine men tried to explain the trade to the dockmaster—they stretched things a little and said wild grouse were the size of turkeys and that for every caribou leaving, two hundred pounds of live birds would be returned. No deal. Heldenbrand tried to call the prime minister of the province. He wasn't home. A storm was coming and there was only a half-ton of lichen in the trucks to feed the caribou. If they missed this ferry, the project was in serious trouble.

They went to the ferryboat captain and pleaded their case. He went to the dockmaster and said he wasn't leaving port without the caribou. The dockside struggle lasted four hours before the ferry set sail with two huge trailers lashed to the deck. In each trailer were twelve tranquilized woodland caribou.

The convoy continued through the small towns of Nova Scotia, most of which have town squares. The people were listening to their radios and waiting in the squares for the trucks. The Mainers paused in acknowledgement as the Canadians wished godspeed with the caribou and then expressed sympathy to the Americans for the loss of their president. John Kennedy had just been shot. It took four days to get from Victoria Lake to Millinocket, the staging area for the transplant of the caribou to the top of Mount Katahdin. The crew was unprepared for the spectacle in Millinocket.

THE NORTH-CENTRAL Maine mill town was alive with media and military. News of the caribou expedition had spread fast and the regional press flocked. Somebody thought it would be a good idea if the Navy tested out its new jet helicopters at high altitude by airlifting the caribou from Millinocket to the tableland atop the mile-high mountain. Margaret Chase Smith, then on the Senate Armed Forces Committee, made a few calls and the Navy appeared in Millinocket, "battle ready."

It was very cold the next morning and nothing the Navy had would start, including the trucks with the special helicopter-starting units on them. "The whole town had to get out to get the Navy going," Heldenbrand remembers. When the first helicopter finally hovered above the gravel pit where the caribou were corralled, it raised up a first-class gravel storm and blew down one of the two big barn doors on the enclosure.

"Here we were with the entire caribou herd staring right at us from five feet away with no doors on the corral," Heldenbrand says. "We had to form a human wall to keep them in."

The Navy got fifteen caribou to the top of Mount Katahdin before "some clouds" came in and they abandoned the operation. "The Navy crept back in disarray," Heldenbrand says. "If that had been a war, we would have been in real trouble."

The remainder of the herd was loaded into trucks and station wagons and hauled to Roaring Brook Campground a few miles from the summit. There, the caribou's blindfolds were taken off and they were released into a gauntlet of men urging them to follow the contours up to the tableland to join their brethren. Man's direct role in the operation was over.

There were sightings of caribou in Baxter Park for about three years. Then there were none. Nobody knows exactly what happened to all of them. Heldenbrand says eventually it was learned at least six were poached by those who considered it "a badge of honor" to kill an endangered species. "One guy actually claimed to have shot two," Heldenbrand says.

Other caribou may have succumbed to the notorious "brainworm" parasite. It is carried by the whitetail deer which spreads it through its feces. It is not fatal for deer, but, in the caribou, the parasite works its way up the spinal column into the animal's brain with fatal results.

It is also possible that the two caribou groups never got back together. They were mature caribou with their migratory instincts developed. They may have tried to go home. In any case, the 1963 caribou transplant failed.

BY 1986, HELDENBRAND and several others involved in the '63 caribou transfer were ready to try again, but this time it would be different. A major difference was that this relocation project would be private, not governmental. The people most interested formed the nonprofit Caribou Transplant Corporation and set about raising money. They needed it: Helicopters for the capture in Newfoundland would cost about a thousand dollars an hour.

Then Maine's commissioner of conservation, Richard Anderson, an enthusiast for the project and a proven money raiser, went to work. This would be a grassroots appeal: How often can you bring something back from the past? Wouldn't you like to support one of the major animal reintroduction efforts of this century in North America? Money came in, but not enough. The eleventh hour was coming with the fall migration and the closing in of winter.

The project's savior was a man who had never seen a caribou, but loved animals. Horace Hildreth, Sr., a former Maine governor, donated fifty thousand dollars late in the campaign and, according to Anderson, "saved the day."

There was a debate about where to locate the caribou. Baxter State Park was rejected. Offshore islands, especially those without whitetail deer, were considered. Allen Island in Muscongus Bay was one, but the Warden Service said it could not guarantee protection of the caribou there. There was some concern that "the guys from Port Clyde would come over to the island with dogs on their boats. They'd let their dogs off on one side of the island and navigate to the other side. The dogs would drive the caribou into the ocean and then the guys would throw lines onto their antlers and hook them onto their lobster hoists and onto the deck." Poaching, maritime-style.

Several factors affected the private group's decision to accept the University of Maine's offer of a fifteen-acre pen in Orono. First, the caribou could be guarded there. Poaching would be unlikely. Second, the University campus would be an ideal place to conduct research on parasites. Everybody understood the threat of the brainworm to the caribou's chance of survival.

A third and telling factor was that, by confining the caribou to a pen, a seed herd could be created. Starting in 1988 or '89, groups of a dozen or so caribou born in Orono would be relocated in various promising locations around Maine. "We're only going to spend the interest this time," Heldenbrand says. "We will colonize, not form just one settlement."

The collection of the caribou in Newfoundland, which began December 5, 1986, also was different from the one in 1963. This time, the media wanted in on the roundup. Almost a hundred representatives of news organizations all across North America, including teams from all major U.S. television networks and the Canadian Broadcasting Corporation, travelled to the isolated province.

This time, the caribou were not selected from a large herd in a corral, but tracked down by two helicopters (each in turn being tracked by helicopters full of photographers and reporters). In winds up to sixty mph, the helicopters hovered above the swiftly moving herds of caribou while wildlife biologists "cut out" promising animals with dart guns.

The ferry waiting in Port aux Basque was a new one called the *Caribou*. The dockmaster presented no hurdles, but the weather did. A savage storm raked the waters between Newfoundland and Nova Scotia, producing waves of up to sixty feet. Heldenbrand said the bow of the ferry was so badly damaged by the mauling waves that the vessel was out of commission for two months. He observed that the passengers "were scared to death" but that the caribou, riding in the hold and not on the deck, had the best ride in the ship.

Twenty-seven caribou were captured and transported in 1986. Two does died on the journey, which featured a second major blizzard before the trucks arrived in Orono after the twelve-hundred-mile trip. Three more died in Orono, one on Christmas Eve. The dead caribou immediately became the subject of intense research, especially for parasites, at the university.

The rest seemed to adapt well, aided not only by the availability of a wide variety of lichen but by a pelletized feed developed especially for the caribou by Blue Seal Feed, a Bangor company which went to the University of Alaska for help with the formula.

The transplant corporation sponsored a series of open houses at the pens in December. Nine thousand people came in a period of twelve hours stretched over four Saturdays. "We outdrew the football team," says Heldenbrand.

The first caribou births came in the spring of 1987, and each seemed a cause for public jubilation. Everybody in the state seemed to be talking about the baby caribou being

## CARIBOU: THE EVOLUTION

born in Orono. As of mid-June, a dozen calves had wobbled their way into the hearts of Maine people. In late June, another open house was held and an estimated six to eight thousand people showed up on one weekend.

It will be years, maybe decades, before we know whether caribou can live in Maine again. But already Ladd Heldenbrand is dreaming: "I envision the day when a family will come up from Philadelphia or some place like that and stop at a publicity bureau. They'll pay ten bucks for a viewing permit and then they'll hire a guide to take them around Maine to see the wildlife. And they will see caribou.

"At no time is the hunting of the caribou envisioned."